AF504379

HIDING IN PLAIN SIGHT... VOLUME II

Hiding In Plain Sight... Volume II

We are the galaxy far far away and you never know who, or what is going to stop by for a visit.

SCOTT GEAREN

Scott Gearen LLC

Copyright © 2024 by Scott Gearen

All rights reserved. No part of this book may be reproduced in any manner whatsoever without written permission except in the case of brief quotations embodied in critical articles and reviews.

First Printing, 2024

FOREWORD

"The photographs in this book do not appear to be "photoshopped". They are enhanced to emphasize the fine details that would not otherwise become apparent.

Mr. Gearen presented me with a "PowerPoint" presentation indicating the "craft" seemed to have advanced stealth and cloaking capabilities. It also seemed able to develop some type of "wormhole" or opening which it eventually stabilized and entered prior to disappearing.

The data and photographs presented to me seemed to be authentic and worthy of further scientific investigation by a civilian scientific body, the military or by an Intelligence Community program".

Based on my personal experiences and knowledge that there are strange and unidentified objects in the sky around us, I highly recommend this book to anyone with an interest in seeing something strange and unidentified, which I believe has not been seen before.

Sabrina Robb, Captain, USAF (Ret)

PREFACE

Why a Volume II?

*Plain and simple answer is: **the pictures**.*

The pictures used in the original "Hiding In Plain Sight" are 100 percent real, and unaltered, but they could have been presented better. There were no filters or other photographic enhancements applied to the pictures included in the first book. Since then, the pictures have been further analyzed, not just by me, but by media experts, through the use of filters, and AI software to reveal details never before seen.

After the first edition of "Hiding In Plain Sight" was published, I was introduced to several individuals with life time experience in Unidentified Aerial Phenomena. One of these individuals who has an extensive background in analyzing pictures, many direct from satellites, primarily NASA related, as well as other ground and aerial platforms asked if I would share them for verification, they were real, and not a hoax through photoshop or other media trickery.

This individual, who wishes to remain anonymous, used an AI software program to enhance each picture, one that is commonly used by government organizations to enhance evidence in criminal cases. They are 100 percent convinced, what I took pictures of is not made by humans.

Although the object is now much clearer, and details not previously seen are exposed, the mystery of what it is, is still present, and it remains an...

Unidentified Flying Object...

To our Creator
For without creation
Nothing would be possible

EPIGRAPH

On new truths...

"No fact is so obvious that it does not at first produce wonder, Nor so wonderful that it does not eventually yield to belief."

"and therefore in other regions there must be other earths inhabited by different tribes of men and breeds of beasts."

Lucretius Carus, c. 99 – c. 55 BC

CONTENTS

| one |

Bottom Line Up Front

The original "Hiding In Plain Sight", published on 18 August 2020, detailed my sighting from 2019, it included comments and guesses about the object I saw, and the pictures I took of it. The pictures are all 100 percent authentic, they were taken on a Sony a6000 camera, with a 210 mm lens, and have been verified by media experts to be real and unaltered by me or anyone else. The same pictures presented here in Volume II are presented in original form and some have been enhanced with software and filters to bring out the details not previously seen in them.

The following information is a recap of this sighting and actions I took after I took the pictures:

On 30 April 2019, I was in Panama City Beach Florida staying in a hotel located on the beach. My room was on the 22nd floor with a balcony attached that provided me a clear and unobstructed view as far as I could see. During the midmorning I was on the balcony when a flicker of very bright light in the sky caught my eye. Possibly due to my innate curiosity or my background in the United States Air Force, when I see something in the sky, I have always tried to identify it. I quickly realized it wasn't an airplane, or a helicopter and it seemed to behave in a strange manner. Because I didn't know what it was, and its strange behavior I began to take pictures of it. Later when I downloaded the pictures to a laptop and was able to see what I had taken pictures of I realized it was in deed something very strange.

Over the next several weeks I tried to determine what it was I had taken pictures of, I realized it was something I had never seen before. My first thought was it must be a secret project the Department of Defense (DOD) was developing; due to this possibility, and my own feelings of loyalty to the DOD, having been enlisted in the United States Air Force for 22 years, I did not want to share the pictures with anyone. I reached out to an individual in the Air Force, an Intelligence Specialist, and explained to them the situation with the pictures I had taken. They told me to send some of the pictures to them and they would pass them to an analyst that was trained in analyzing pictures taken from satellites and other airborne platforms.

Several weeks later, after prompting from me, I asked what the analyst had determined, and was told the individual was unable to determine what it was I had taken pictures of and was sending the pictures to another agency with better capability to analyze pictures. I never asked which agency they were going to; I was surprised the analyst didn't know what it was but was happy to know that the pictures were going to another agency and I would have an answer soon. Little did I know that answers don't come easily in regards to Unidentified Flying Objects (UFOs).

Again, several weeks later, I asked what the unknown agency had determined about my pictures. I was told they said the object I had taken pictures of was a "balloon, and my camera had an anomaly that caused these weird shapes". With thoughts of "balloons" and "swamp gas" being the official identification applied to many UFOs since the Roswell incident, I decided to go public with these pictures and see what others would say about it. I uploaded several videos to a YouTube account, filed an Unidentified Flying Object report with both MUFON (#103707), and NUFORC (#147831), and decided to publish a book with the pictures: on 18 August 2020 Hiding In Plain Sight was published.

After I submitted my report to MUFON I was contacted by Field Investigator, Sabrina Marie Robb, Captain, USAF (RET), to determine, if possible, what I saw could be identified and if not then come to a best guess conclusion about what I had witnessed. As a The Field Investigator, Sabrina was very thorough, we initially spoke on the phone, and we were able to meet at a later date where I presented the pictures in a Power Point from my laptop. I also provided pictures, which were independently analyzed by others to confirm they were real. The Field Investigator was able to go online and watch the videos I uploaded as well as the multiple websites that had picked up my report. In the end the Field Investigator was not able to identify the object and told me, *"I do NOT believe any of the technology on this craft was invented by us... even though it may now be "ours",* and their best guess was this object was probably a future aircraft or some type of future technology. The Investigator told me they would submit their findings to MUFON Headquarters stating the object was "unknown", and would suggest that it be further investigated by MUFON. This report is identified as #103707. It is unknown if anyone from MUFON did further investigation into this sighting.

The videos and my statement in the NUFORC report were picked up by news-papers, individuals, and organizations with websites such as British Earth and Aerial Mysteries Society (BEAMS) who felt strongly enough about it to say this:

"We suspect that this object may well be what is termed as a 'crypto creature'; these are biological entities that have not yet been officially discovered/ recognised by science.""

"That is not such an outrageous proposal, because, just as many very unusual species of fish are now being discovered in the previously unexplored depths of the world's oceans, 'cryptos' like the one you see here may live most of the time in the still relatively unexplored, upper reaches of our atmosphere... perhaps coming down to 'feed' off the pollution and energies that man emits."

Based on this analysis, along with new reports that some of the unidentified sightings might be "Terrestrial" or even "aquatic" in nature. I believe this object is in part or in whole a biological entity. Possibly a newly discovered species.

This is not the first hypothesis for a "space animal", the "Space Animal Hypothesis" suggests there may be animal lifeforms indigenous to the Earth's atmosphere. In 1948 Arthur Conan Doyle wrote about "saucer-like space animals in his book "The Horror of the Heights", ironically the illustrations for this book look noticeably like the jellyfish shaped object in a video released in 2023, where the video shows something that resembles a jellyfish floating through the air over a military base in Iraq in 2018.

Also, after the book was published, I was introduced to an individual (US Army Retired), who is a recognized expert in the field of digital media, to evaluate my pictures using his professional equipment and software. Several of this expert's comments, after thorough examination and analysis include:

"I do NOT believe these images to be a hoax", I have never seen anything like what these pics show. They are way beyond my attempt to associate these objects with anything I know about. These are super interesting images. No idea what they are. I wish I could help more ".

While I observed the object, I realized there were times when I was unable to see it with just my eyesight, and the only way I could see it was while wearing polarized sunglasses. Since the polarized lens are a filter, I concluded that by applying filters available to me through media tools on my laptop that I might be able to locate the object in pictures that only appeared to be empty blue sky.

After trying several filters, I was able to identify what appears to be a round sphere, and I have included several of those in Volume II that were not included in the first book. My basis for believing the sphere is the object is due to how it was captured in a sphere shape in several still pictures, and in videos taken during the same sighting where it can be seen that the object is indeed changing from shape to sphere to shape.

The pictures presented in Volume II are a mix of the original published pictures, some are enhanced, and some are shown with filters applied to provide additional contrast to them. Thank you for looking, I hope you find this object, still unidentified, to be as interesting and unique as I have.

| **two** |

Unidentified Flying Object

UFO Picture 0986: April 30, 2019: 11:42:12 AM CST
Scott Gearen

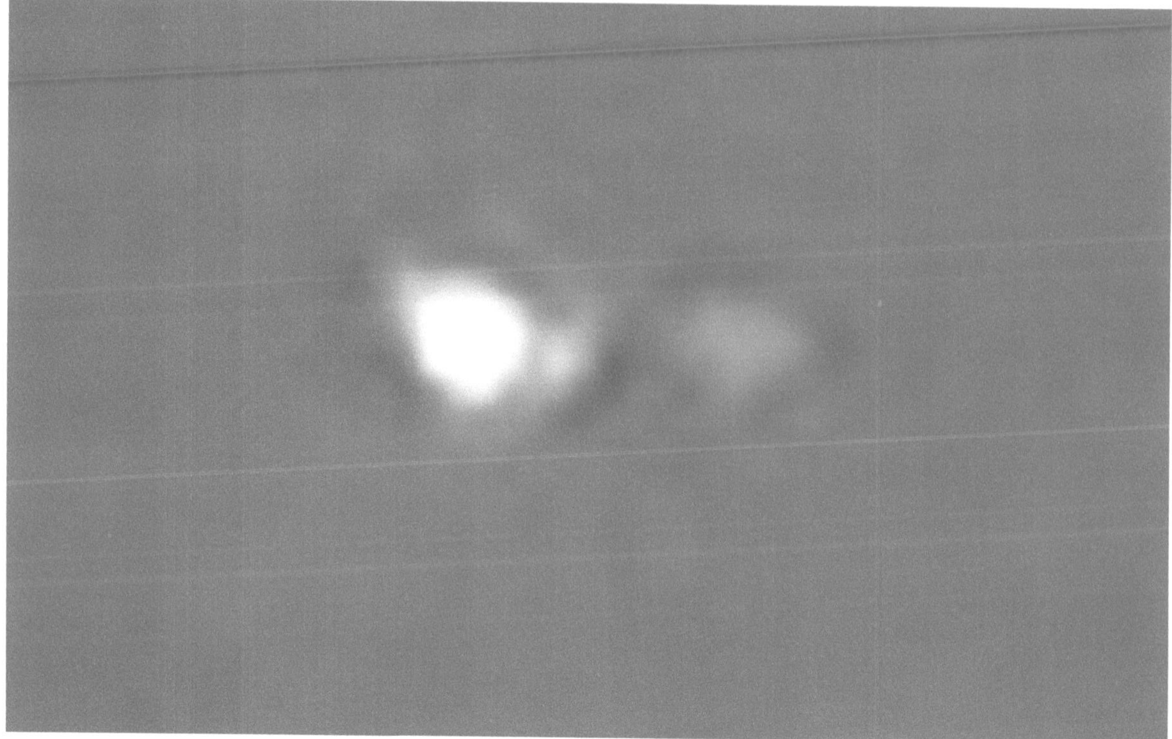

UFO Picture 0986: April 30, 2019: 11:42:12 AM CST
Scott Gearen

UFO Picture 0987: April 30, 2019: 11:42:16 AM CST
Scott Gearen

If you catch a glimpse of something in the sky, a flash of light, a reflection that seems odd to you, don't dismiss it, try to determine if you know what it is, if you cannot identify it, it is an Unidentified Flying Object, a UFO... maybe it is a secret government project or maybe it isn't...

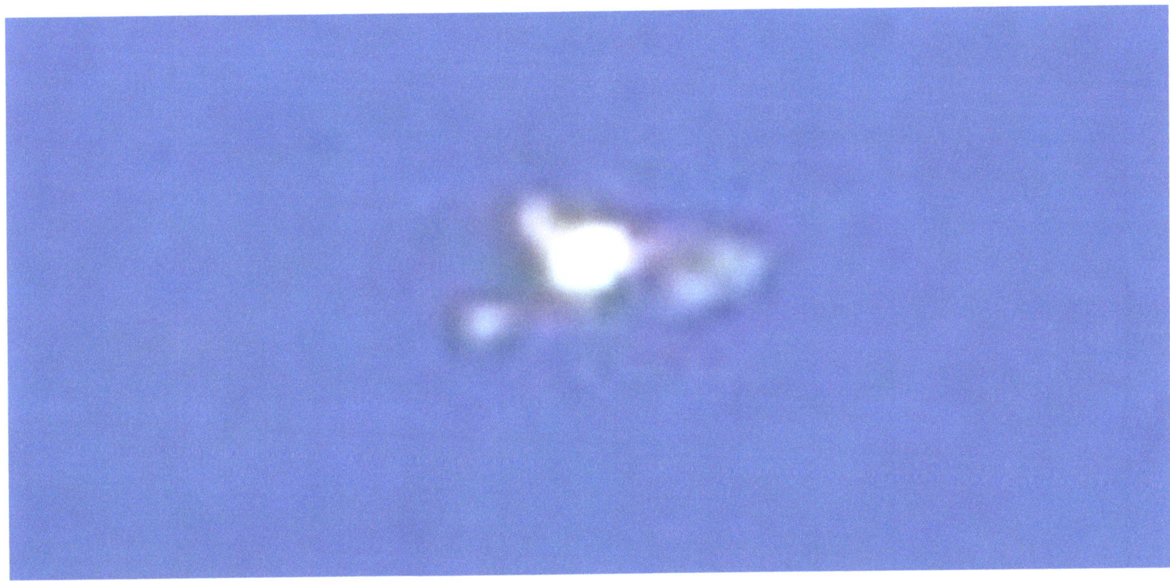

UFO Picture 0987: April 30, 2019: 11:42:16 AM CST
Scott Gearen

UFO Picture 0988: April 30, 2019: 11:42:20 AM CST
Scott Gearen

UFO Picture 0988 (Zeke Filter): April 30, 2019: 11:42:20 AM CST
Scott Gearen

UFO Picture 0992: April 30, 2019: 11:42:34 AM CST
Scott Gearen

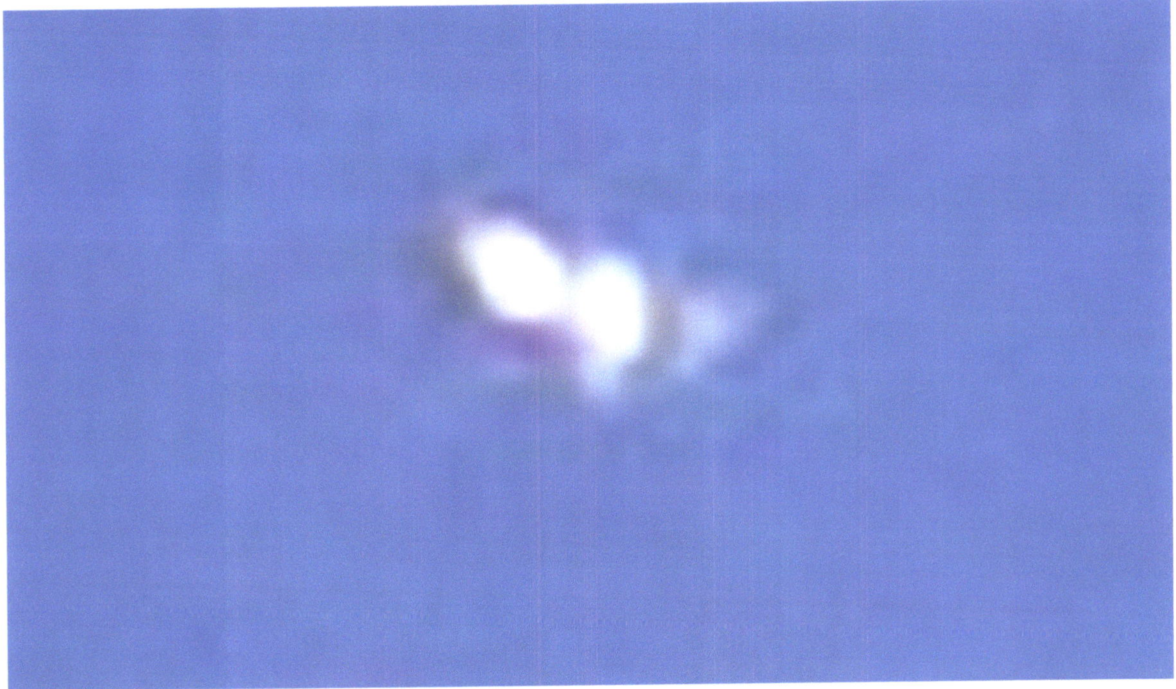

UFO Picture 0992: April 30, 2019: 11:42:34 AM CST
Scott Gearen

UFO Picture 0993: April 30, 2019: 11:42:38 AM CST
Scott Gearen

UFO Picture 0993: April 30, 2019: 11:42:38 AM CST
Scott Gearen

UFO Picture 0994: April 30, 2019: 11:42:40 AM CST
Scott Gearen

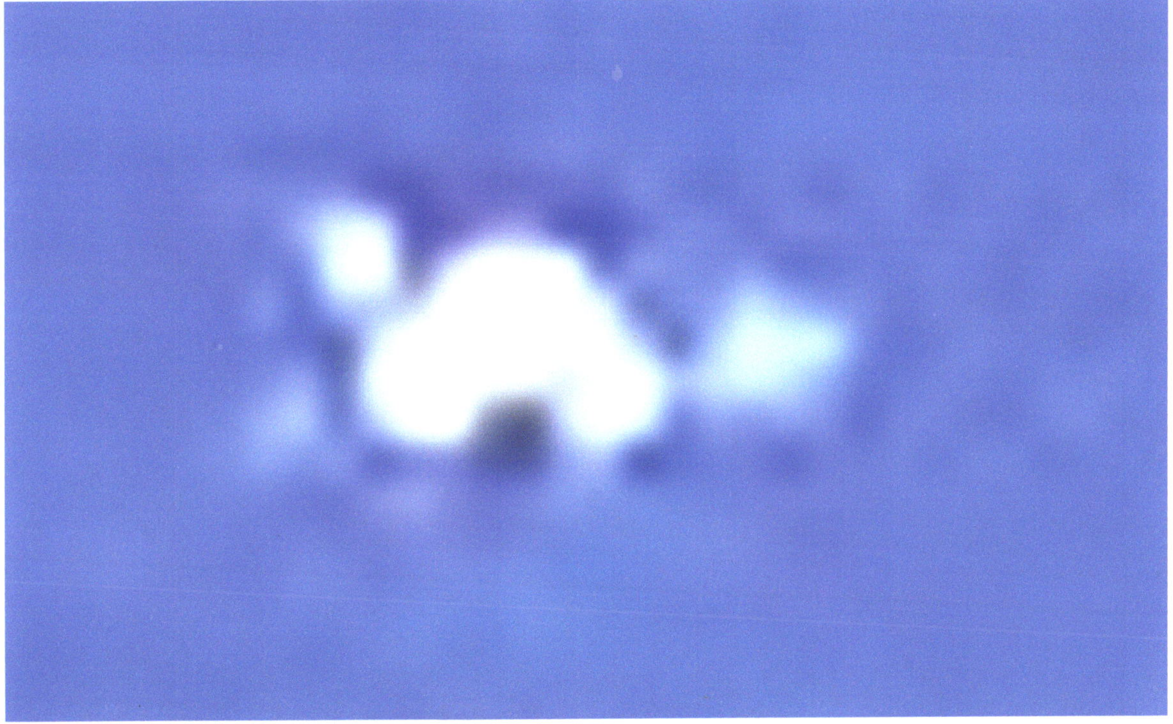

UFO Picture 0994: April 30, 2019: 11:42:40 AM CST
Scott Gearen

UFO Picture 0996: April 30, 2019: 11:42:56 AM CST
Scott Gearen

This is the original picture top and zoomed in and cropped at the bottom. At first, I thought the dark section was a wing and with the "empennage" and horizontal and vertical stabilizer, at the rear, just like a modern airplane… but after closer examination and the enhanced pictures it is easy to see this is much more than just a craft, and more in line with something "biological"…

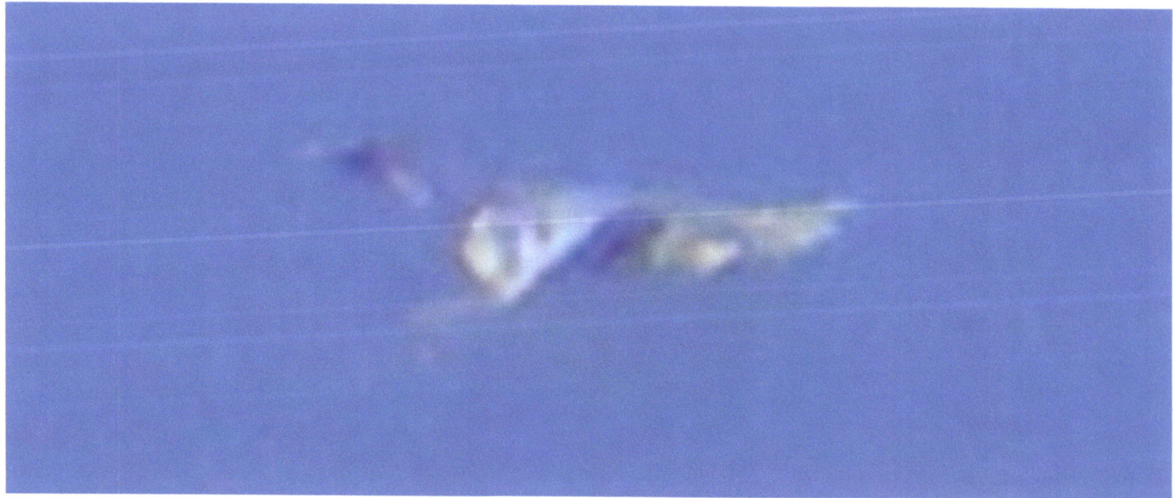

UFO Picture 0996: April 30, 2019: 11:42:56 AM CST
Scott Gearen

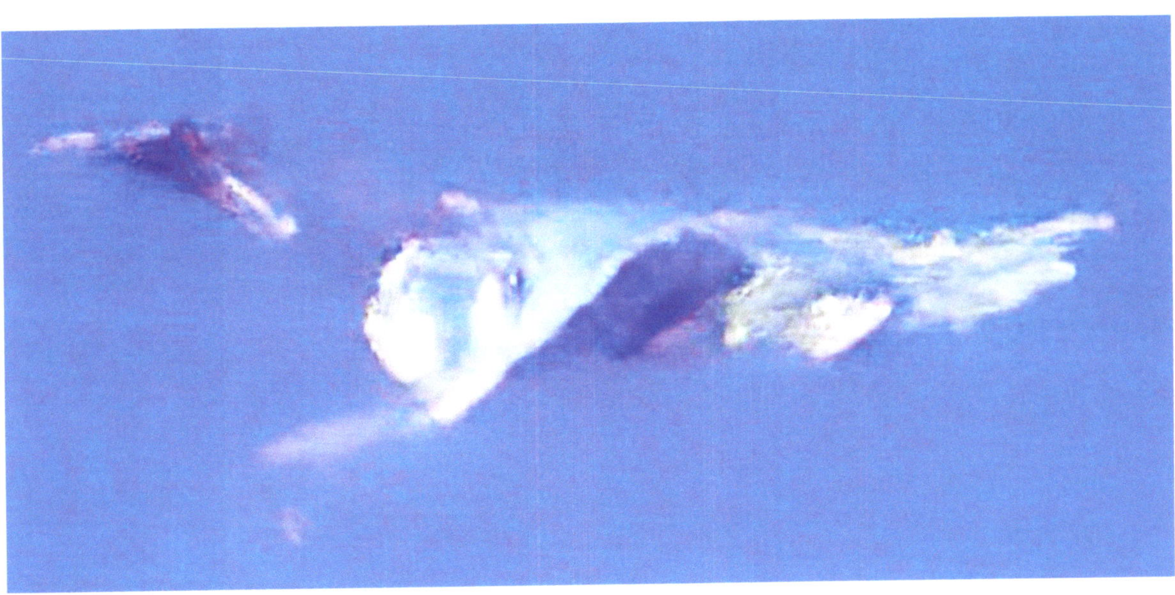

UFO Picture 0996: April 30, 2019: 11:42:56 AM CST
Scott Gearen

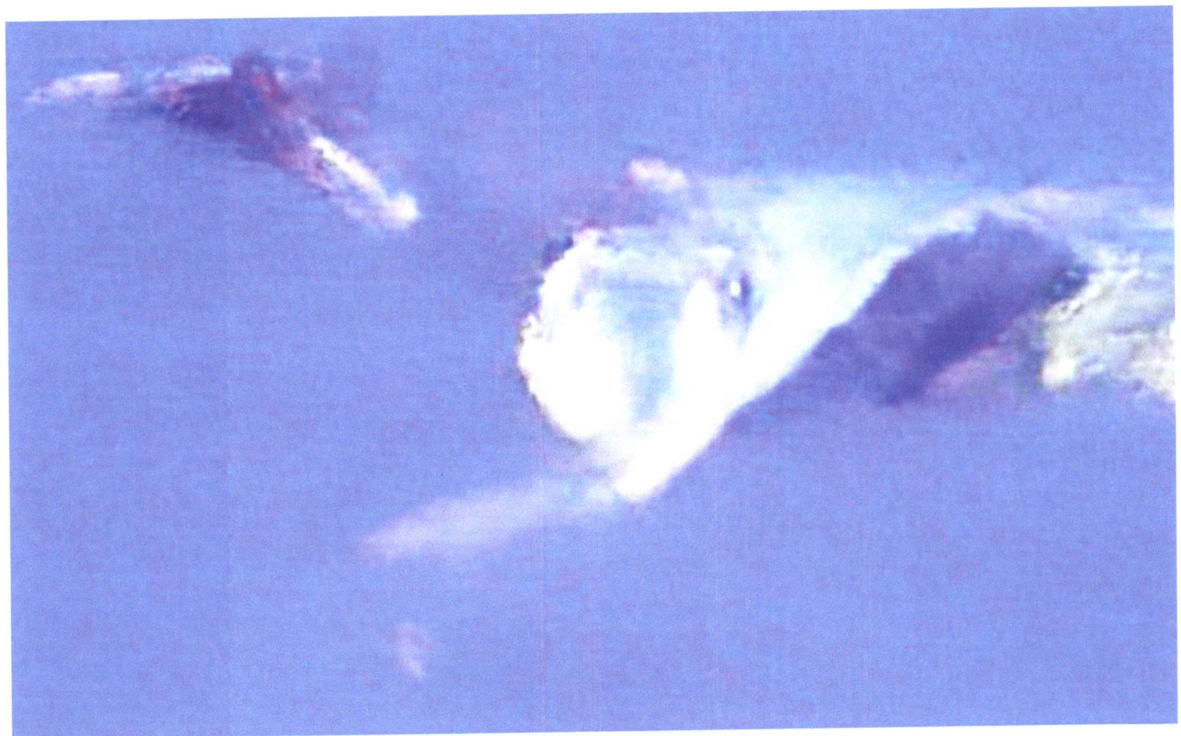

UFO Picture 0996 (Head, Mouth, and Eyes): April 30, 2019: 11:42:56 AM CST
Scott Gearen

Have ever see a Tarpon, a large silver scaled fish found in oceans around the world, if you have you can see the resemblance, a large open mouth as it is gulping down a meal. This was my first thought when I saw this enhanced picture, there is even a small pink colored object that is either going in the mouth or is coming out. Is this a Cryptologic Creature, in the atmosphere feeding on something or is this a mouth brooder spitting out its young? Are those eyes on each side of the mouth and another next to the dark section in the middle. Is it alive?

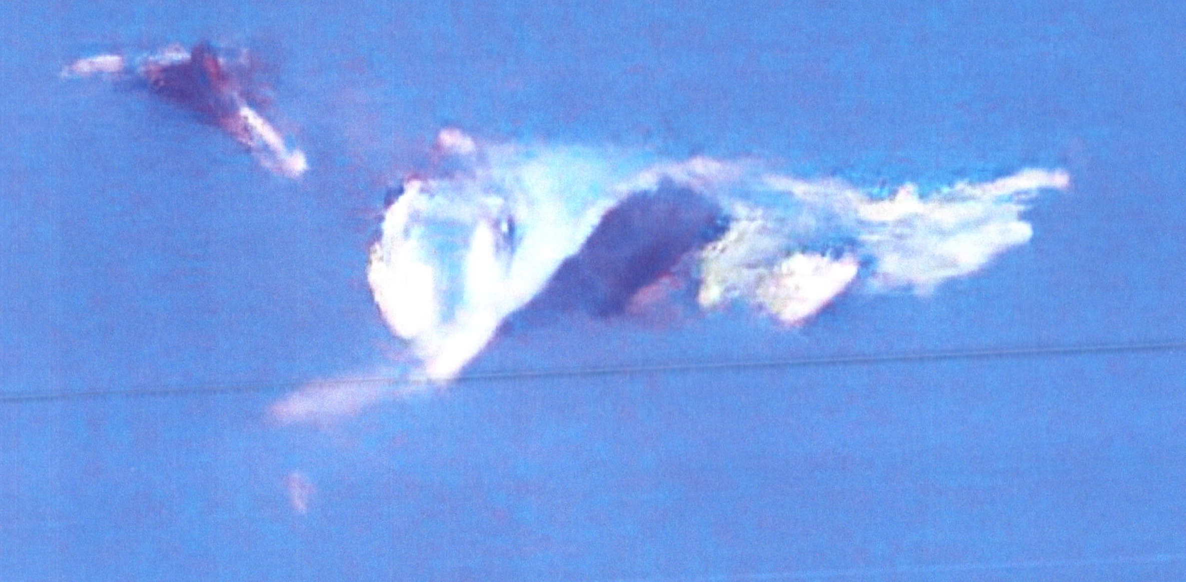

UFO Picture 0997: (Zeke Filter): April 30, 2019: 11:43:04 AM CST
Scott Gearen

The object would change into a circular shape and disappear so fast my eyes couldn't see it, but the camera caught it. I knew it was there, and was able to locate it after viewing with filters.

UFO Picture 0997: April 30, 2019: 11:43:04 CST
Scott Gearen

UFO Picture 0998: April 30, 2019: 11:43:08 AM CST
Scott Gearen

What is the object doing; is it coming or going? Why does it change shape? Could this object be moving in and out of a dimension we cannot see and are unaware of? Could this be a worm hole, dimension to dimension or into another galaxy...

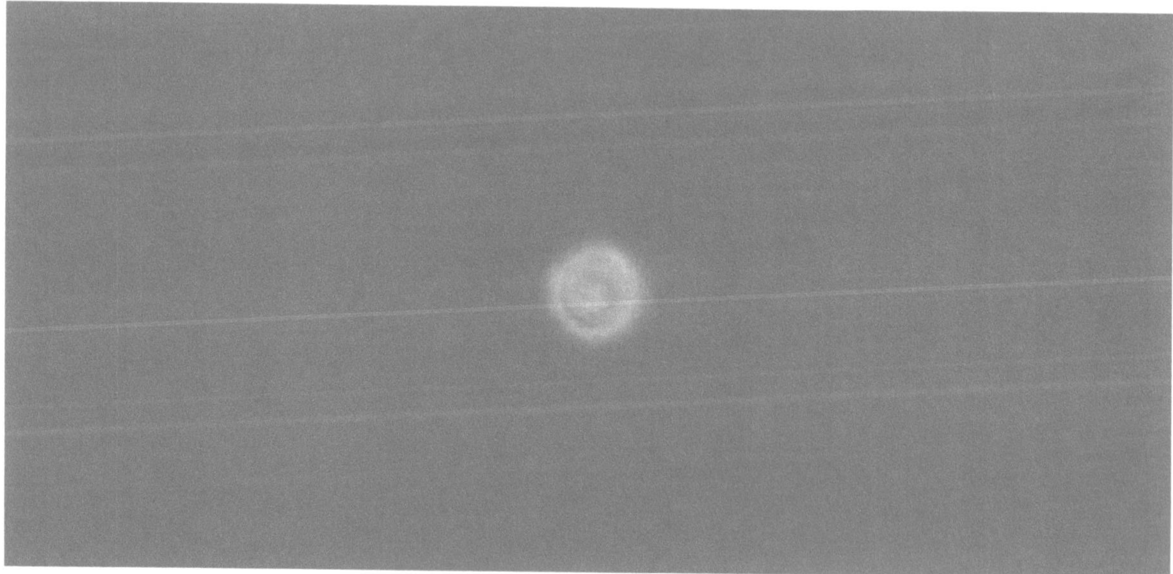

UFO Picture 0998: April 30, 2019: 11:43:08 AM CST
Scott Gearen

UFO Picture 1000 (Zeke Filter): April 30, 2019: 11:43:18 CST
Scott Gearen

Another picture where the object can only be seen through use of filters. There are several dark spots on the lens, they are in the same location in every picture, they are used as a "landmark" when the object is located with filters, just like in this picture. There were many pictures I took of the object and could only locate it in some but not all pictures with the filters available to me.

UFO Picture 1000: April 30, 2019: 11:43:18 CST
Scott Gearen

UFO Picture 1001: April 30, 2019: 11:43:22 AM CST
Scott Gearen

At first, I thought the picture was blurry, and it is, but not due to my camera. The *object is blurry*, the camera is focused. Many pictures, taken by me and by others are blurry like this, it isn't the camera, it is the object. The object might be coming into view or disappearing, there is energy being emitted around it, maybe a gas that distorts the air around it as the object transforms itself.

UFO Picture 1001: April 30, 2019: 11:43:22 AM CST
Scott Gearen

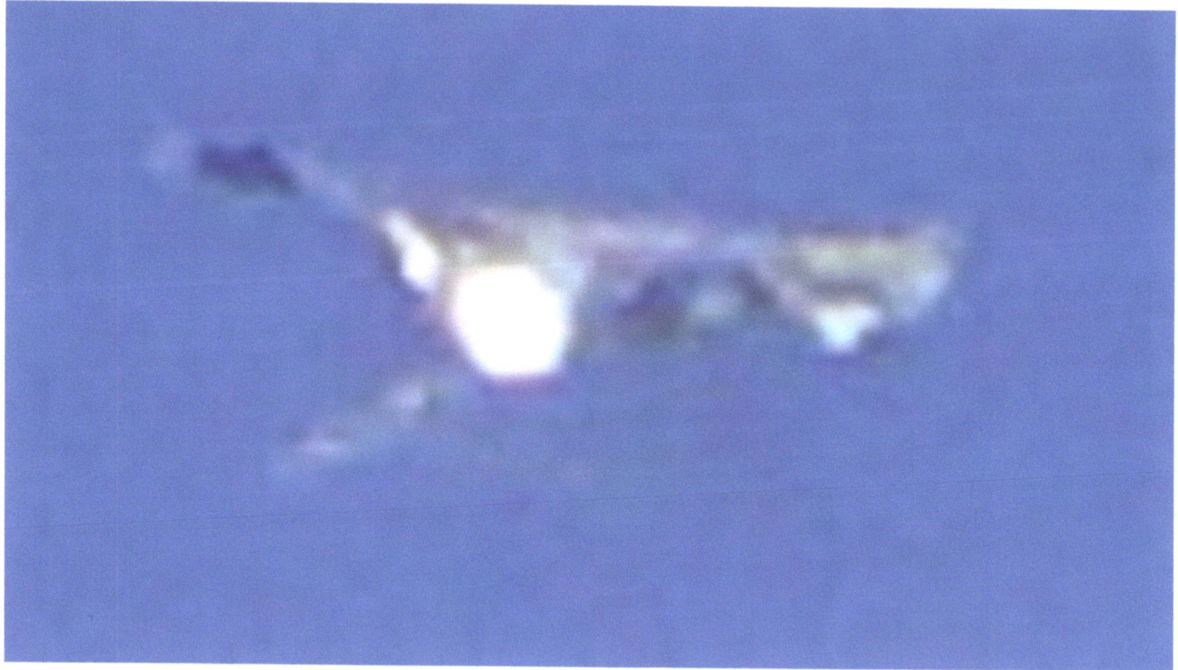

UFO Picture 1002: April 30, 2019: 11:43:26 AM CST
Scott Gearen

In just four seconds the object can be seen more clearly. Slightly different shape but similar to others. The bottom picture is the original and the top it is zoomed in to see details. The bright area looks like an engine, but it isn't, and continues to increase and decrease in size over the entire object as it changes shape, until it covers it entirely, and then it disappears. Is this Anti-Gravity?

UFO Picture 1002: April 30, 2019: 11:43:26 CST
Scott Gearen

UFO Picture 1002: April 30, 2019: 11:43:26 AM CST
Scott Gearen

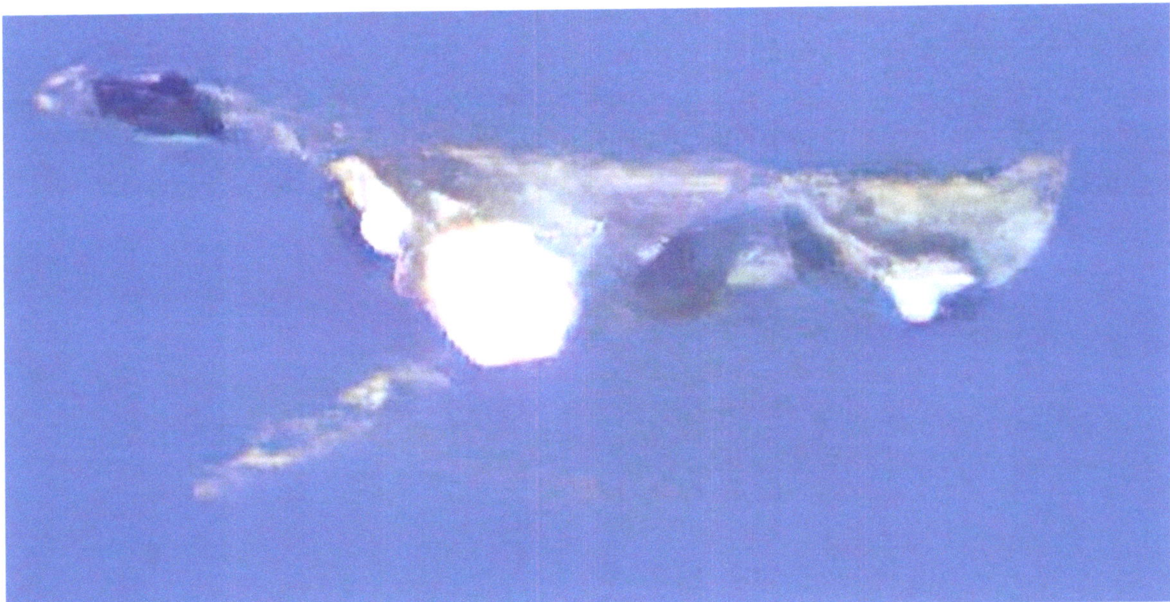

UFO Picture 1002: April 30, 2019: 11:43:26 AM CST
Scott Gearen

Picture is enhanced and we can clearly see the object appears unstable, more like plasma, but something continues to hold the plasma within defined borders, just as our skin holds everything inside our bodies, we grow, we change, but we always stay inside our defined border... our skin, just like the skin of every living creature we are aware of. This object is not moving horizontally the way a modern aircraft would to stay aloft, yet it stays aloft. Are those wings?

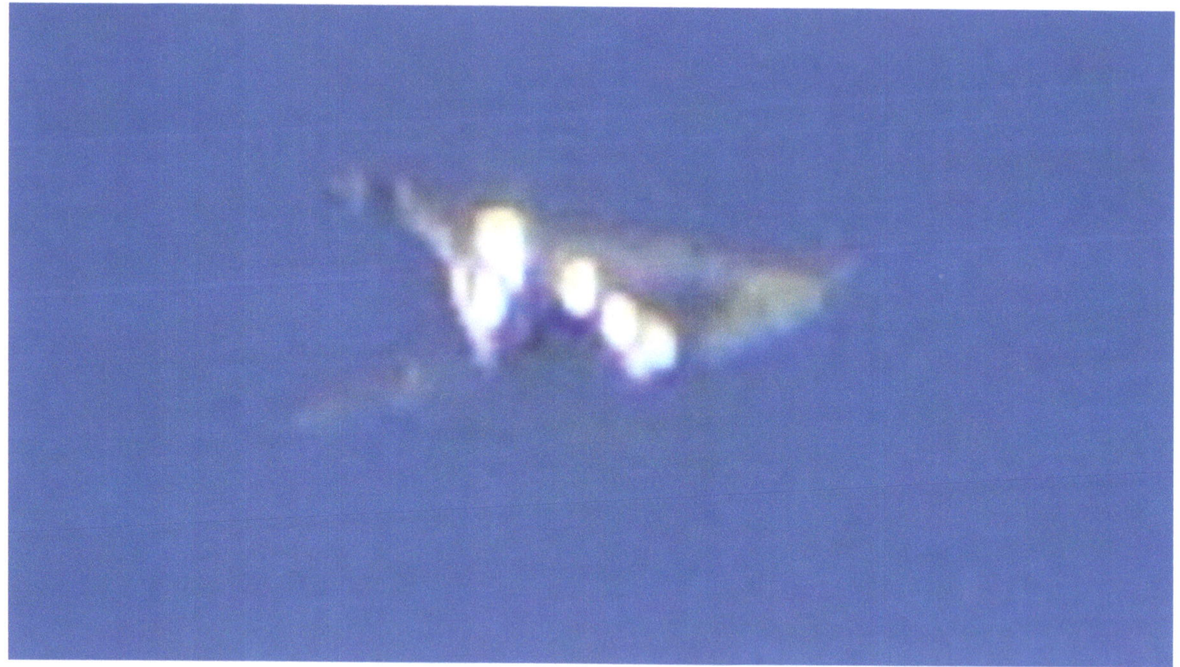

UFO Picture 1006: April 30, 2019: 11:44:18 AM CST
Scott Gearen

The large bright area has changed into several smaller bright areas, and moved into different areas of the object, it doesn't appear to be a specific pattern, maybe it is, but it is a continuous cycle where the object quickly changes shape and each time the bright areas continue to change locations, increase and decrease in size with different levels of brightness, until it becomes a circular ball of light, disappears and then reappears in a different shape, and the cycle continues to repeat.

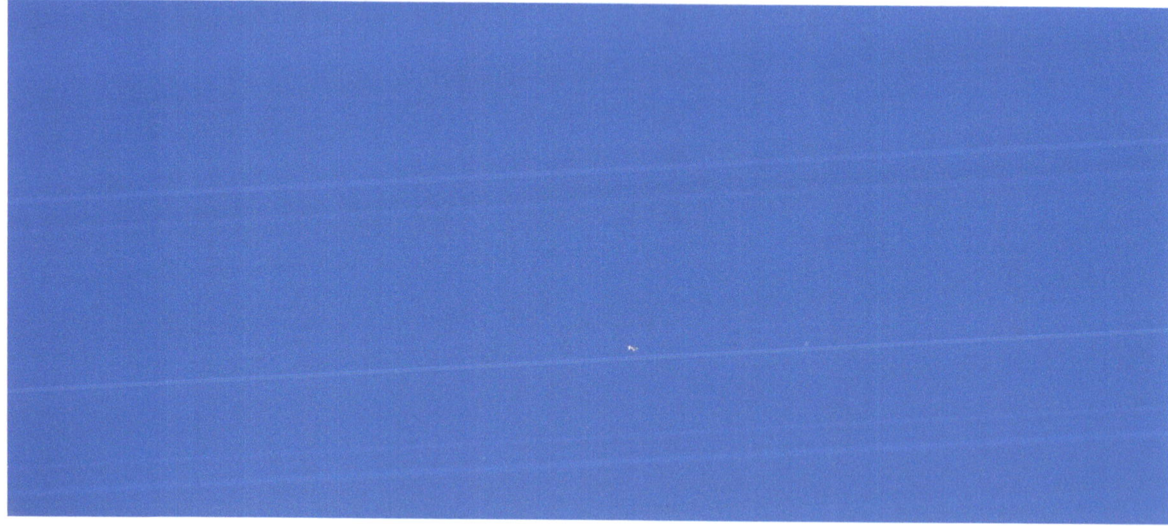

UFO Picture 1006: April 30, 2019: 11:44:18 AM CST
Scott Gearen

UFO Picture 1006 (Sunscreen Filter): April 30, 2019: 11:44:18 AM CST
Scott Gearen

Both pictures, top and bottom have been enhanced with a different filter applied to each. Filters allow you to see the object differently than with just eyesight alone. Sometimes elements not seen become visible, but not always, in this case the underside of the near "wing" can be seen as it appears to be turned up, and some areas are slightly clearer. The dark circular section in the middle of the object is distinct, and appears in multiple pictures, maybe it is an internal capsule for occupants... Is this a biological craft capable of transporting occupants?

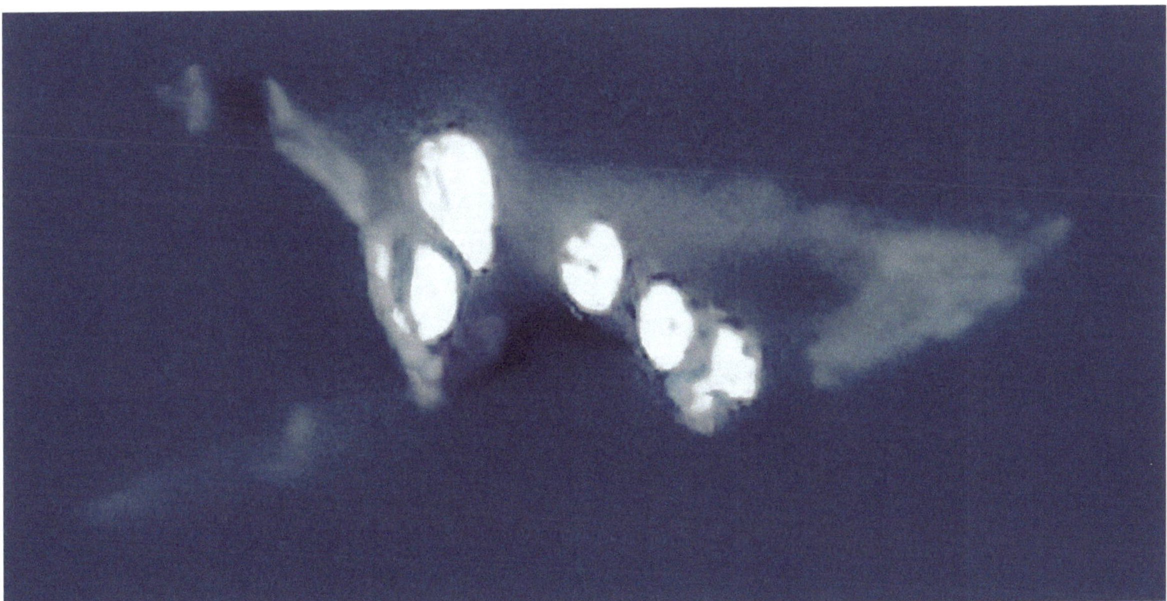

UFO Picture 1006 (Denim Filter): April 30, 2019: 11:44:18 AM CST
Scott Gearen

UFO Picture 1007: April 30, 2019: 11:44:24 AM CST
Scott Gearen

Once the object reaches a certain level of intensity it appears to spin at an extremely high rate of speed, it quickly fades from view and will then reappear, always in a different shape. The sequence is repeated over and over and is presented in the series of pictures in this book. This picture has captured the object in what appears to be two sections, a very bright sphere and a circular ring, similar to a smoke ring, behind it. Could this object have moved so fast the camera was able to capture it in two places, or is the ring a portal the object utilizes to travel between dimensions?

UFO Picture 1007: April 30, 2019: 11:44:24 AM CST
Scott Gearen

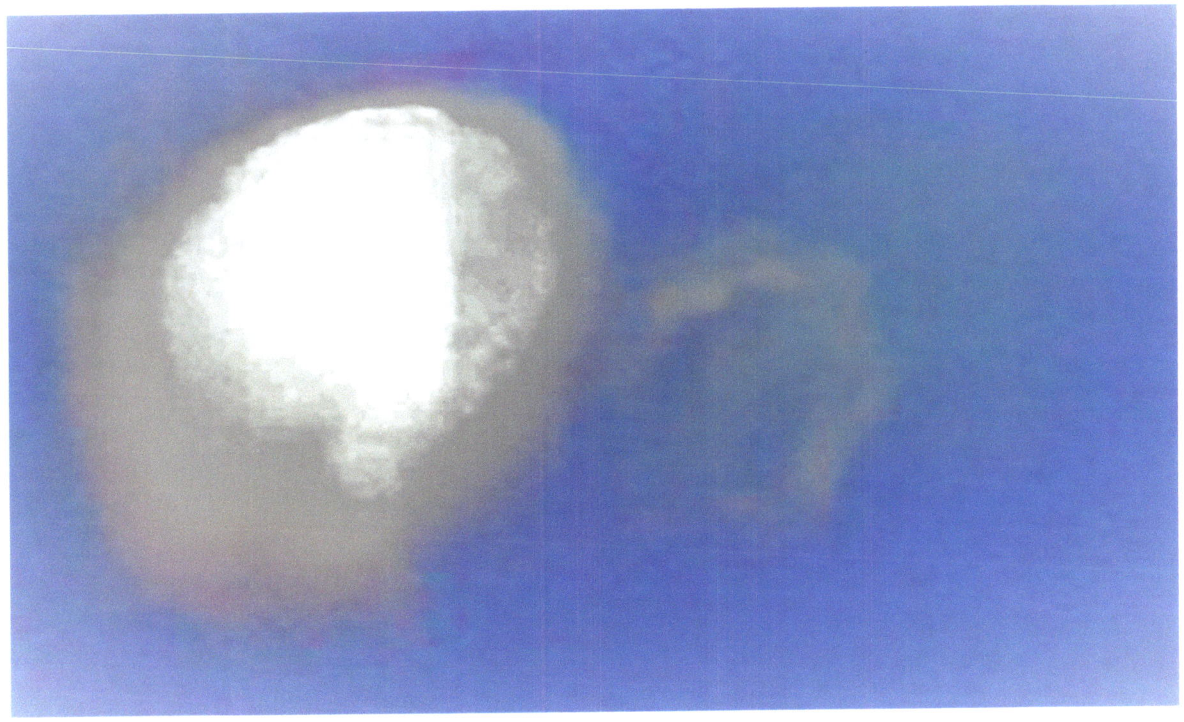

UFO Picture 1007 (Icarus Filter): April 30, 2019: 11:44:24 AM CST
Scott Gearen

UFO Picture 1007 (Sunscreen Filter): April 30, 2019: 11:44:24 AM CST
Scott Gearen

Both filters used show a depth to the "ring" section, there really is no way of knowing what the object is doing with or without the filters, but the object is not stagnant as it remains aloft, it is clearly going through transformations for a reason. What that reason is remains unknown. Further analysis of these pictures continues to reveal new questions without answers.

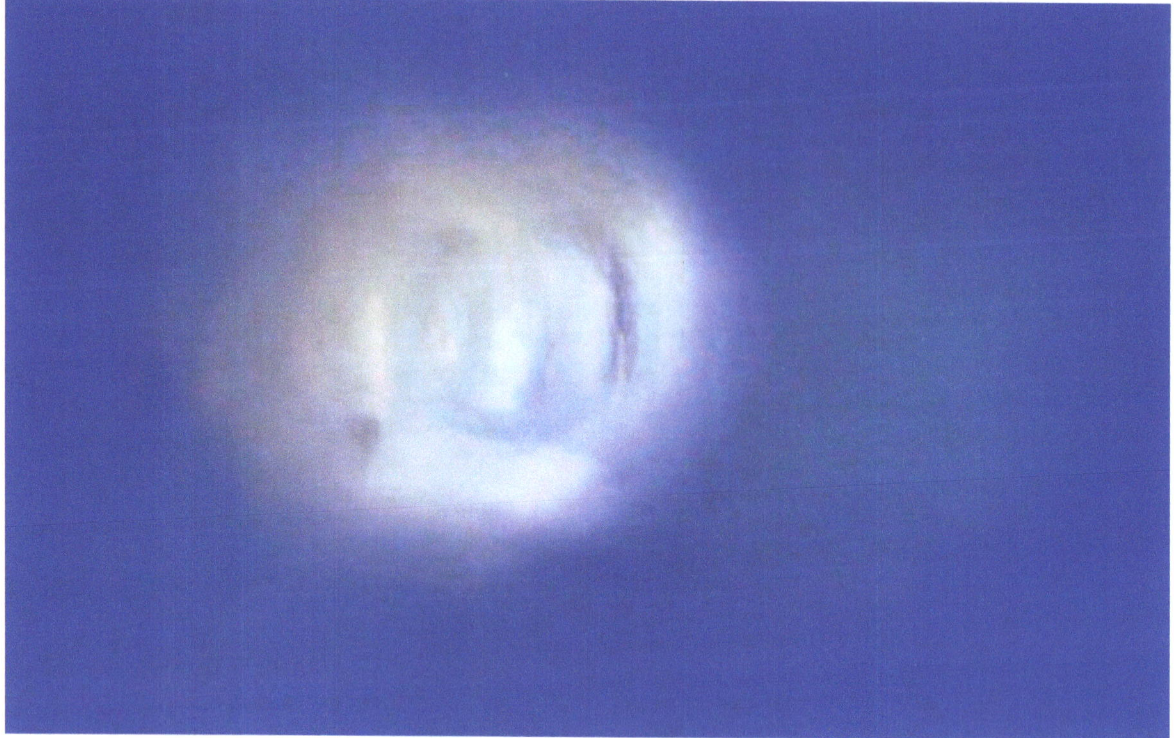

UFO Picture 1008: April 30, 2019: 11:44:32 AM CST
Scott Gearen

The bright sphere is now showing a more circular pattern rather than a solid bright object. Rings and spaces appear showing it is not solid, the track of the circular rings, similar to contrails as seen behind modern jet aircraft, but rather than a straight-line contrail they are wound tightly and appear to be moving in this circular pattern at a high rate of speed, possibly slowing down as a new shape will appear within a few seconds.

UFO Picture 1008: April 30, 2019: 11:44:32 AM CST
Scott Gearen

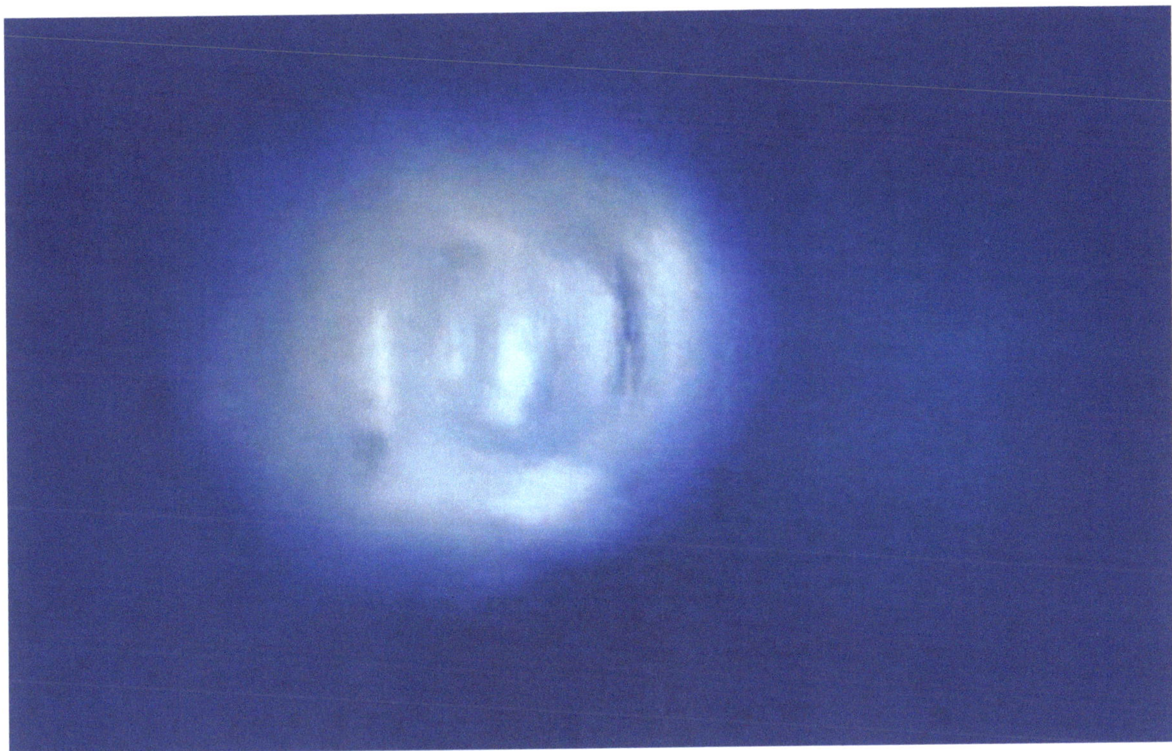

UFO Picture 1008 (Mercury Filter): April 30, 2019: 11:44:32 AM CST
Scott Gearen

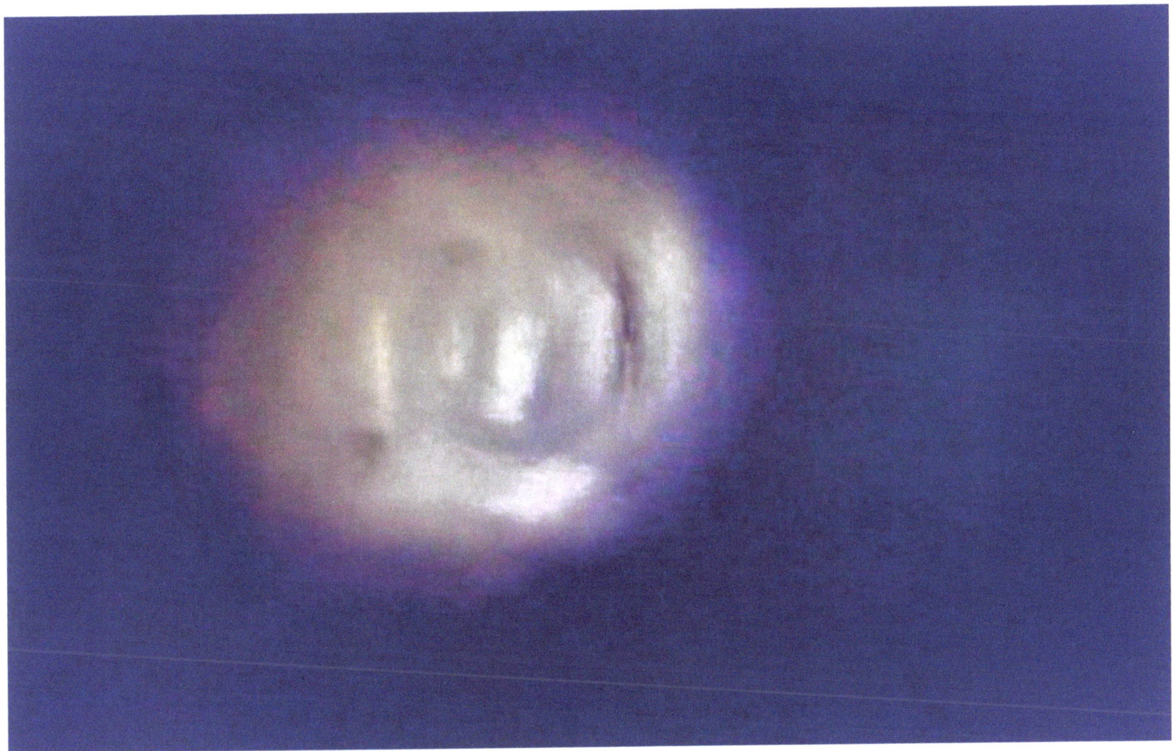

UFO Picture 1008 (Zeke Filter): April 30, 2019: 11:44:32 AM CST
Scott Gearen

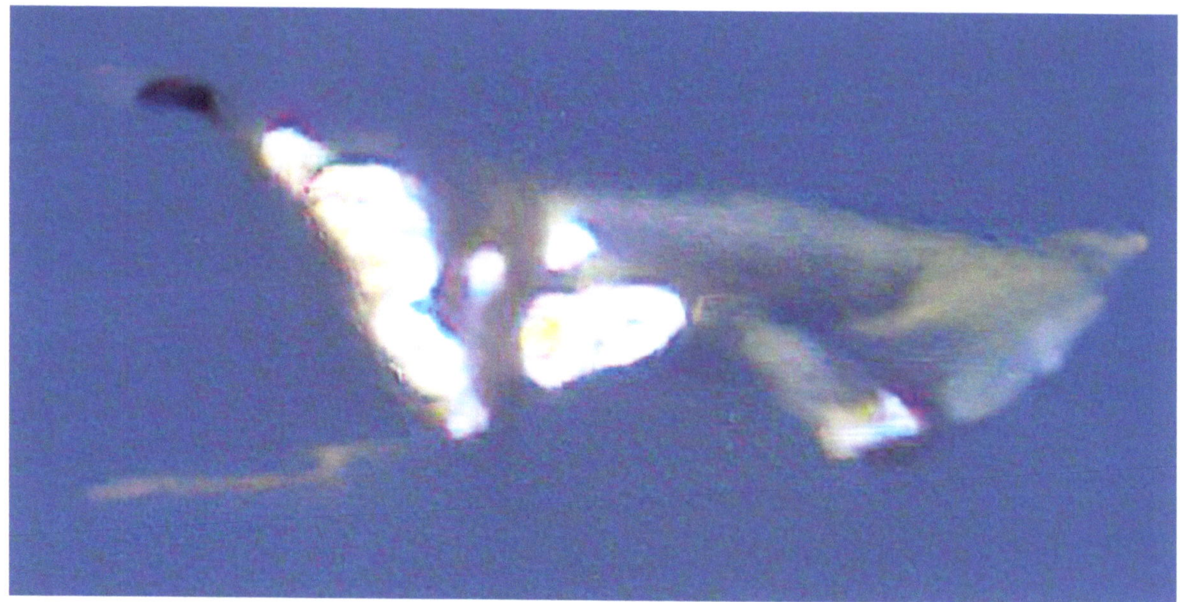

UFO Picture 1009: April 30, 2019: 11:44:38 AM CST
Scott Gearen

A similar but different shaped object emerges from the bright sphere. Bright irregular shaped areas appear spread out over the object, a wing is evident and both the blue underside and the top can be seen. The same rounded nose, which appears to be the front is present along with what appears to be an eye on each side, and the dark area reappears in the middle area of the object. The object appears to be both fluid and solid at the same time. Is this a craft of unknown origin, a secret classified project, or a biological entity just discovered? Whatever it is, where did it come from?

UFO Picture 1009: April 30, 2019: 11:44:38 AM CST
Scott Gearen

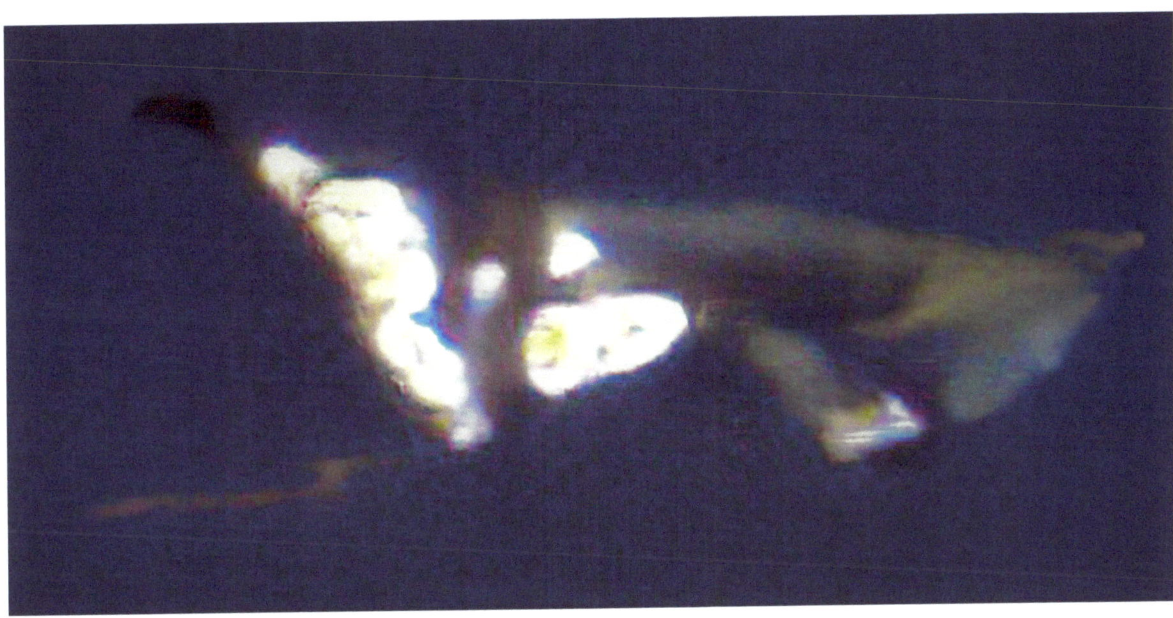

UFO Picture 1009 (Sunscreen Filter): April 30, 2019: 11:44:38 AM CST
Scott Gearen

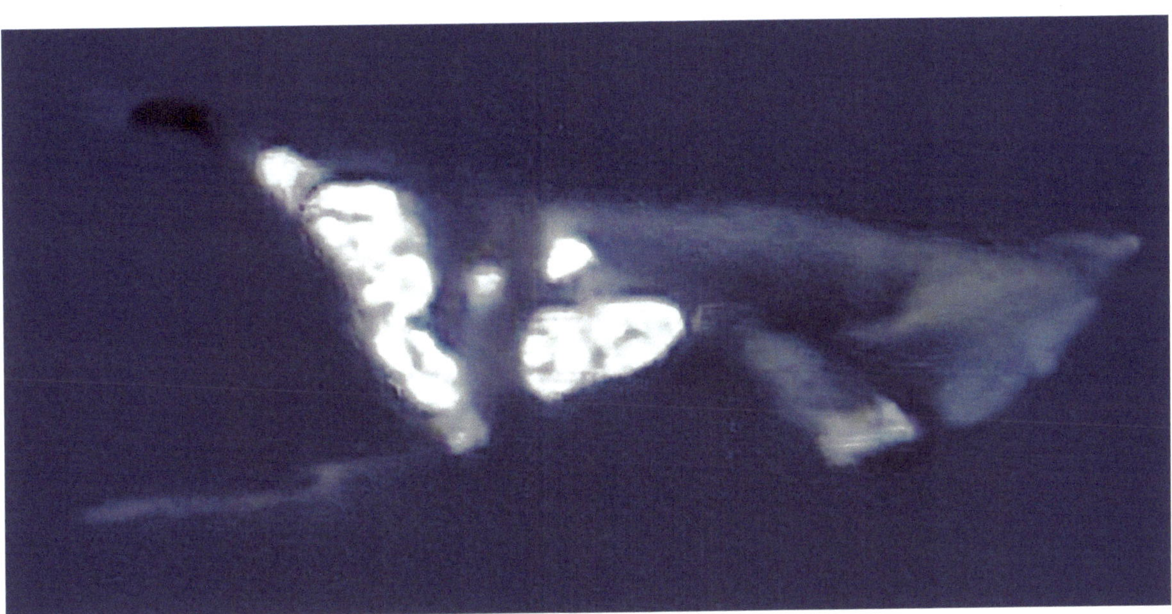

UFO Picture 1009 (Denim Filter): April 30, 2019: 11:44:38 AM CST
Scott Gearen

How can an object like this not be seen? Surely with the advanced radars that protect and defend the airspace above the United States, the Defense Department would see and identify this object. The explanations from people in authoritative positions that this is a balloon is no longer believed by anyone. By human definition of a living being, this object appears to be alive. Is this a biological creature that has evolved naturally, or is this a craft that has been created by Non-Human Intelligence, or possibly used by future Humans returning for a reset?

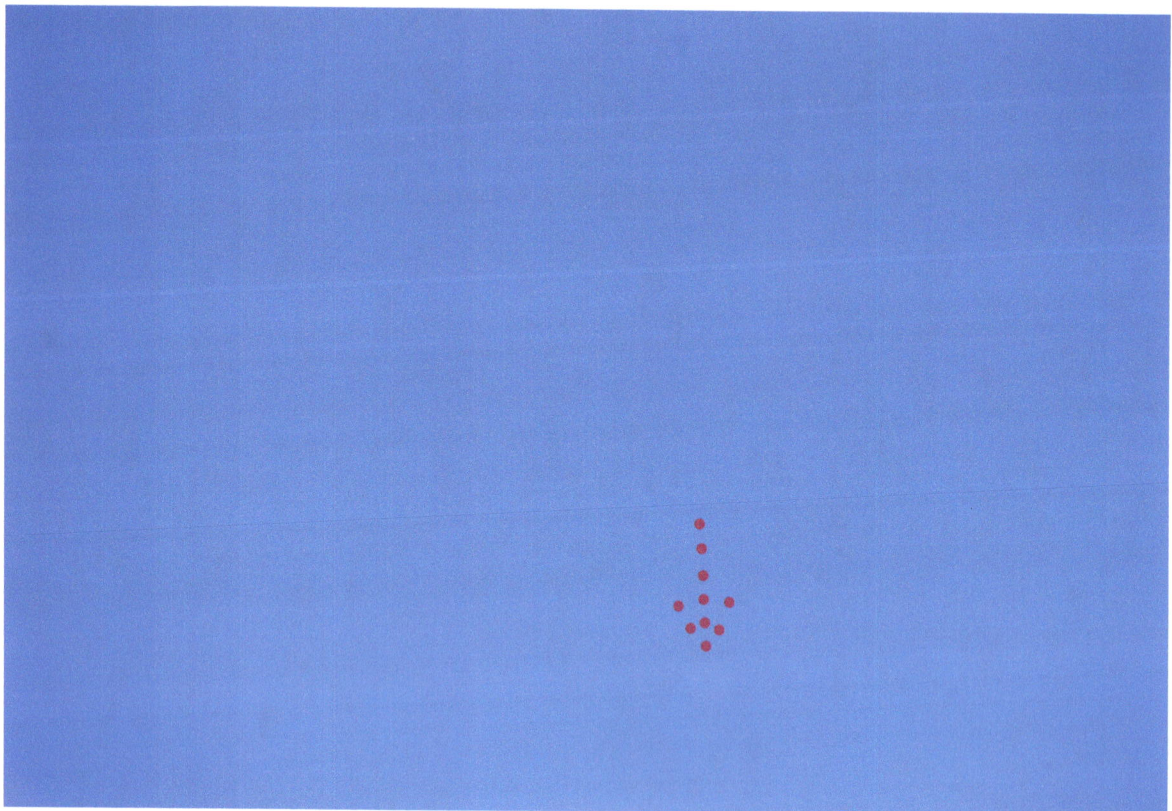

UFO Picture 1011: April 30, 2019: 11:44:54 AM CST
Scott Gearen

Without a picture that can be enlarged, and thoroughly analyzed, this object would never be noticed by anyone, even when looking into the sky. Each time I pressed the shutter button on my camera there was an object in view; the object was changing so quickly it appeared there was nothing in the picture. It was only through continued analysis and use of filters was it located.

UFO Picture 1011 (Zeke Filter): April 30, 2019: 11:44:54 AM CST
Scott Gearen

UFO Picture 1012: April 30, 2019: 11:45:10 AM CST
Scott Gearen

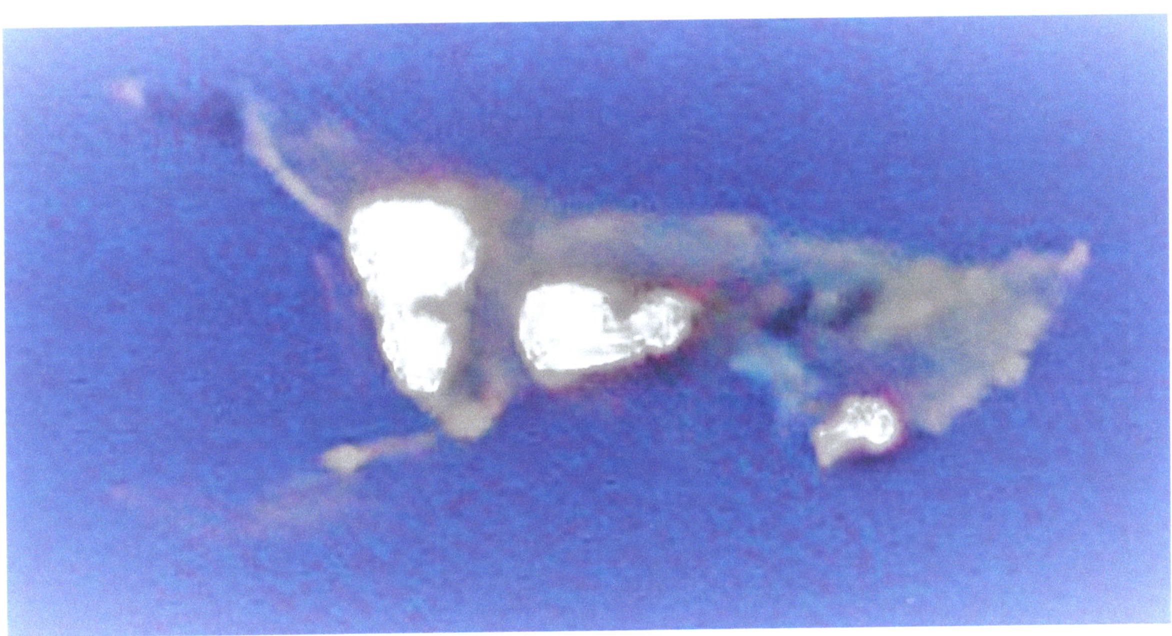

UFO Picture 1012 (Icarus Filter): April 30, 2019: 11:45:10 AM CST
Scott Gearen

Top picture is enhanced and enlarged, it is another similar object that is oriented in the same direction as all of the objects, except for the moments when it is spherical shaped. There appears to be a head section, wings on the sides and tail in the rear, and a dark area in the middle. The bottom picture is enlarged with an Icarus filter applied. In the lower right, what I think may be an "eye", now appears to be more mechanical than natural. Is there another "eye" on the opposite side?

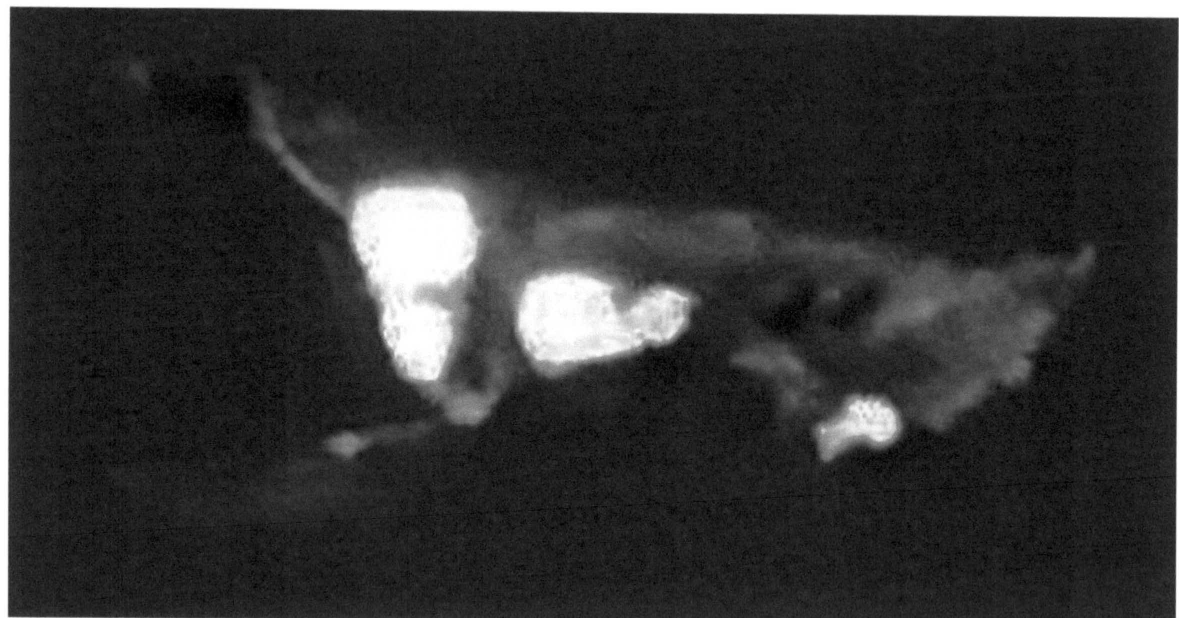

UFO Picture 1012 (Denim Filter): April 30, 2019: 11:45:10 AM CST
Scott Gearen

Denim filter is applied to this picture and enlarged with focus on the "eye" section of this object. It doesn't identify what it is but it does clearly show this is a multi-dimensional object.

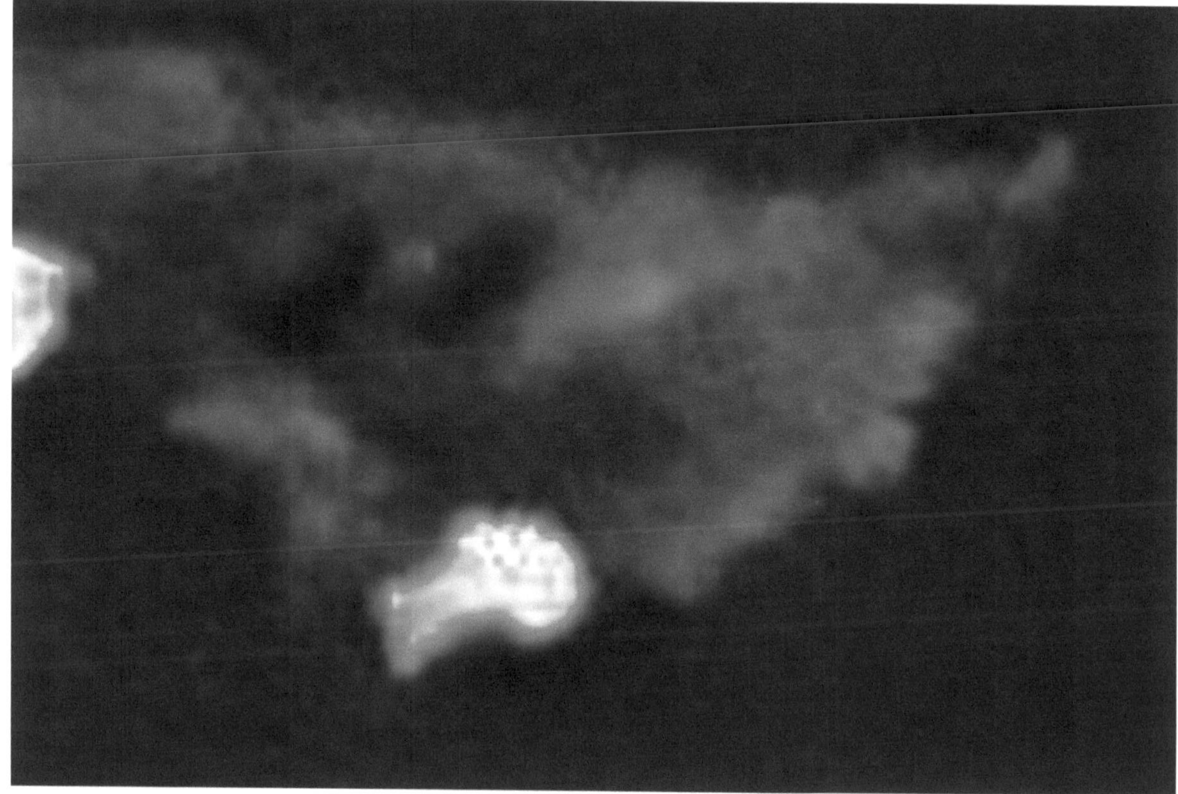

UFO Picture 1012 (Denim Filter): April 30, 2019: 11:45:10 AM CST
Scott Gearen

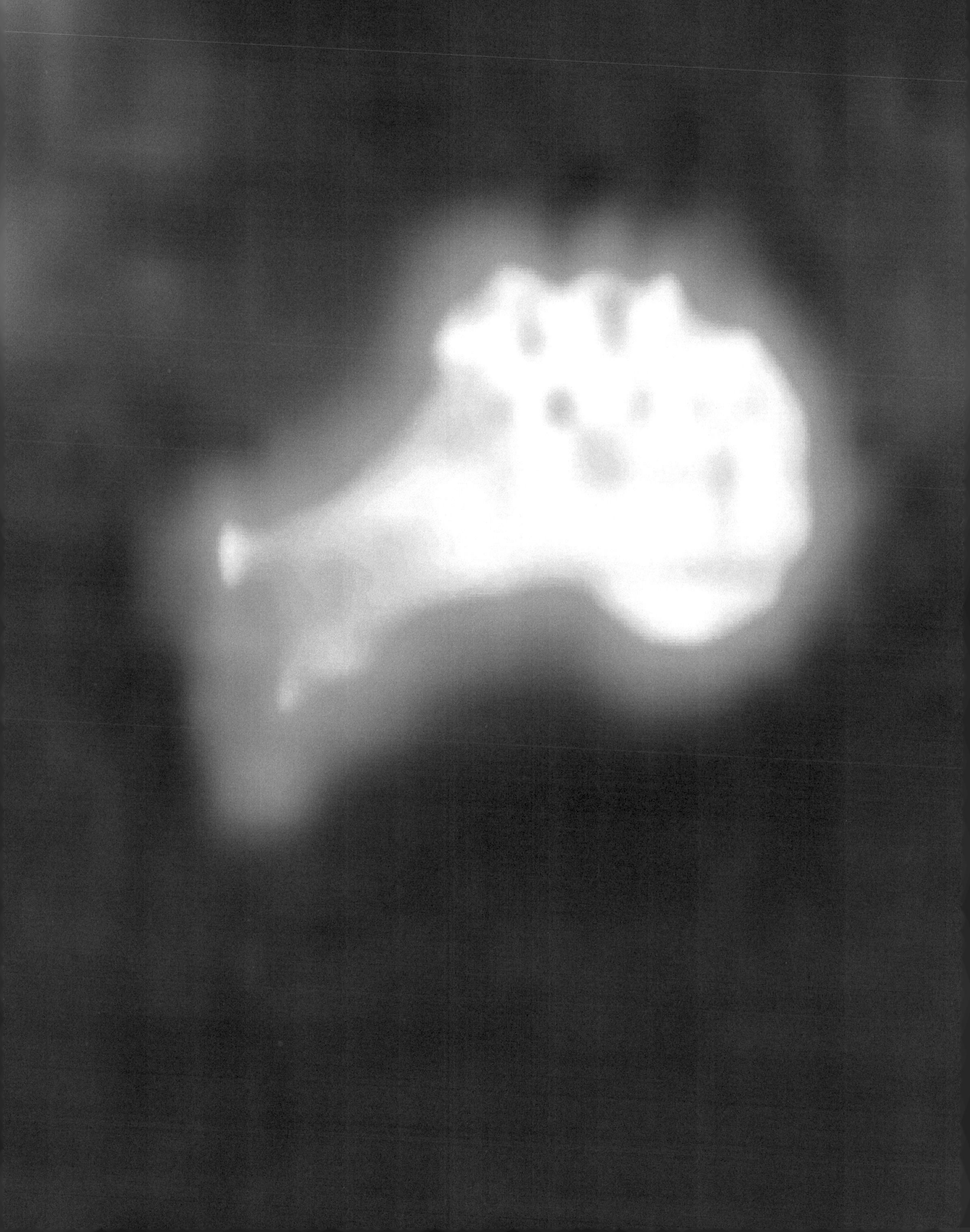

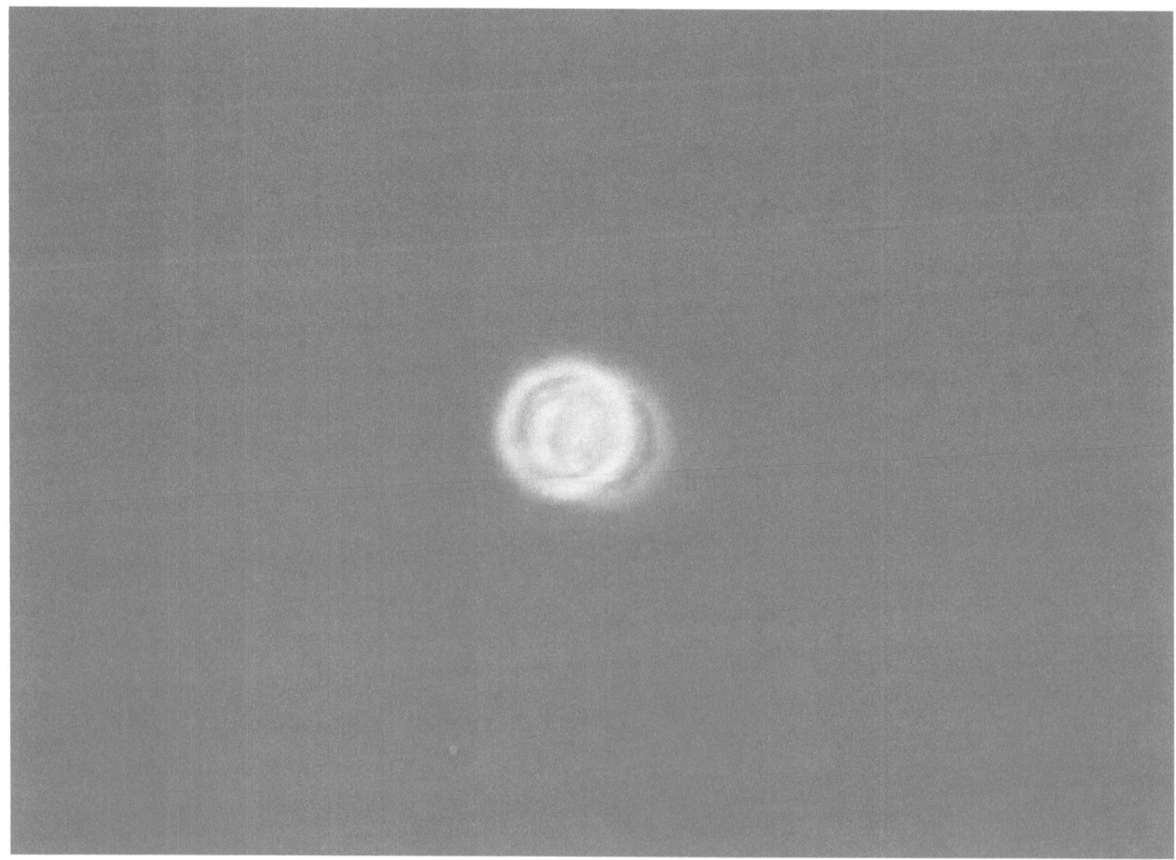

UFO Picture 1013: April 30, 2019: 11:45:46 AM CST
Scott Gearen

Another picture of the object as it transforms, it appears to be spinning at a high rate of speed, but is it possible the object is simultaneously going into or out of another dimension?

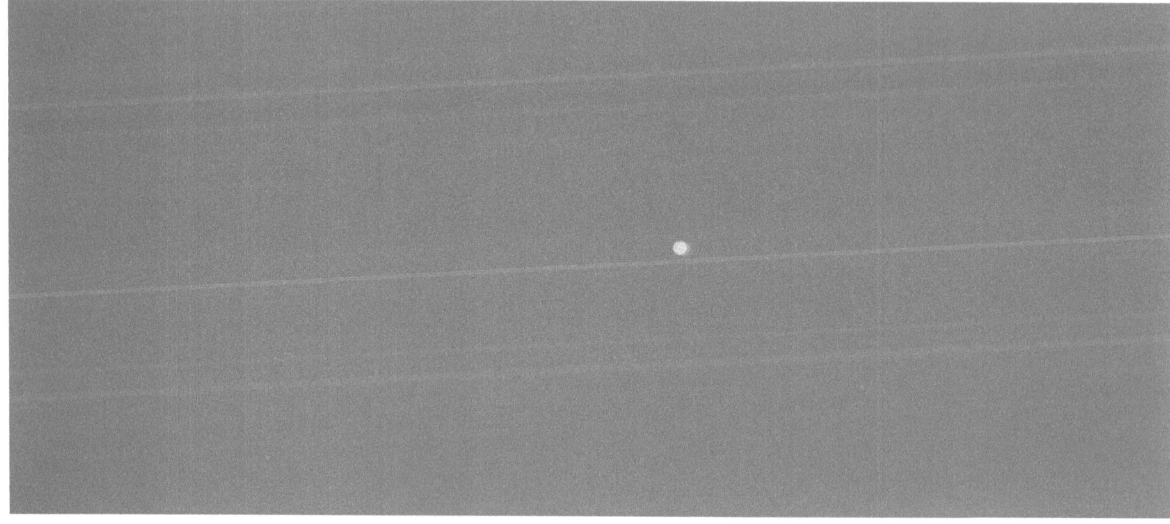

UFO Picture 1013: April 30, 2019: 11:45:46 AM CST
Scott Gearen

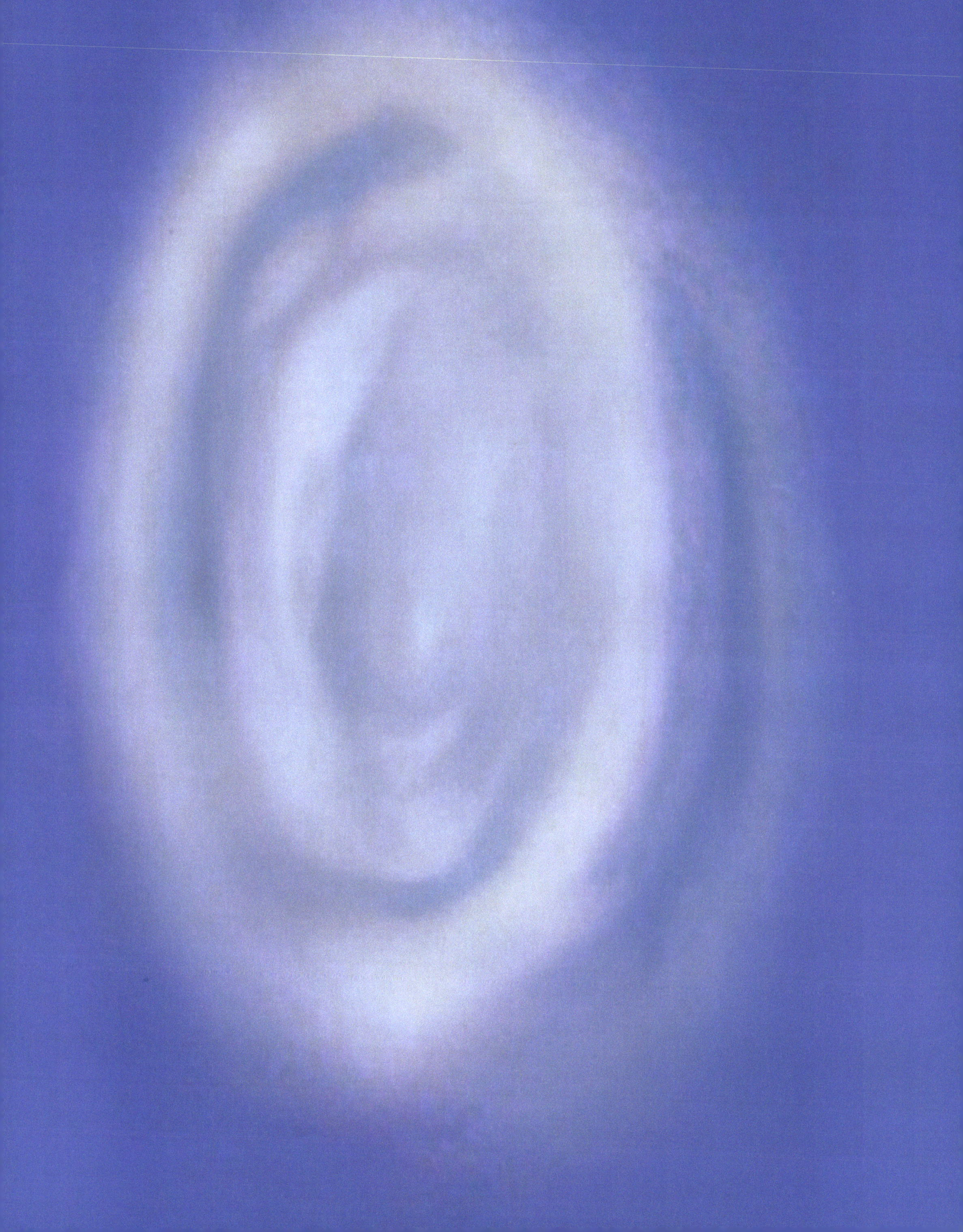

UFO Picture 1014: April 30, 2019: 11:45:52 CST
Scott Gearen

From circular rings this object appears in just six seconds. Without the entire series of pictures it would be easy to dismiss this as just an out of focus picture, but the reality is the picture appears blurry, not because the camera is out of focus, but the object is a blur of motion as it is transforming from one shape to the next. The movement of the object and the energy it creates is making it appear blurry. Some reports claim the UFO splits into two sections, is this what is happening?

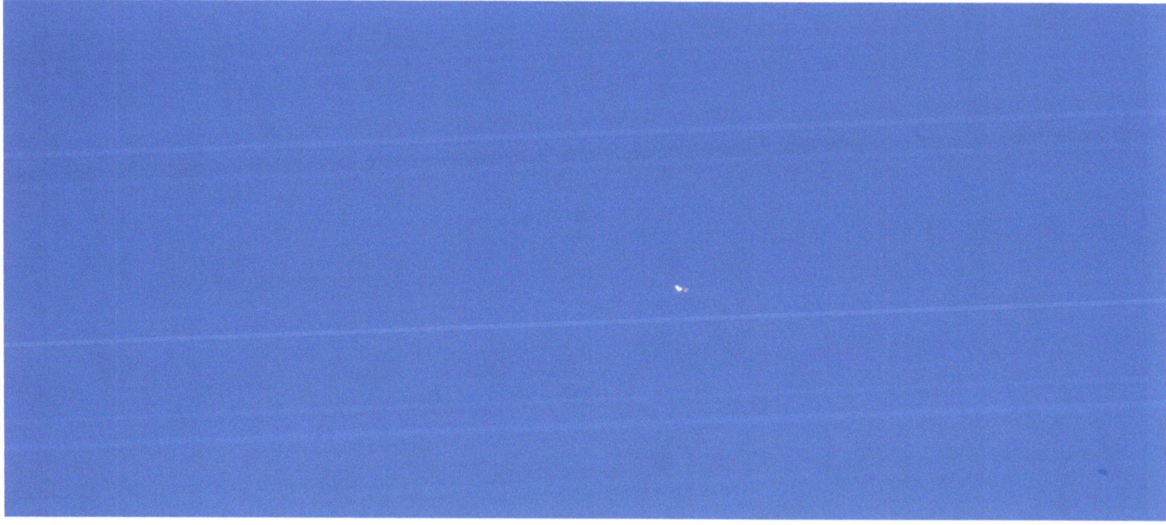

UFO Picture 1014: April 30, 2019: 11:45:52 AM CST
Scott Gearen

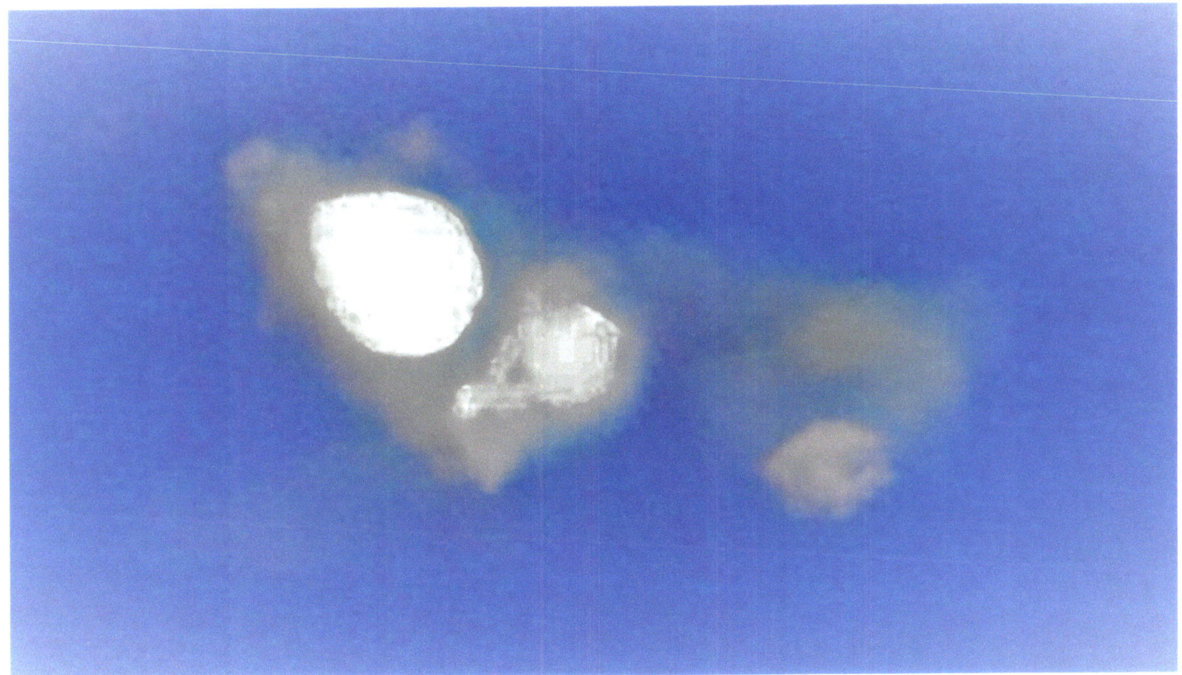

UFO Picture 1014 (Icarus Filter): April 30, 2019: 11:45:52 CST
Scott Gearen

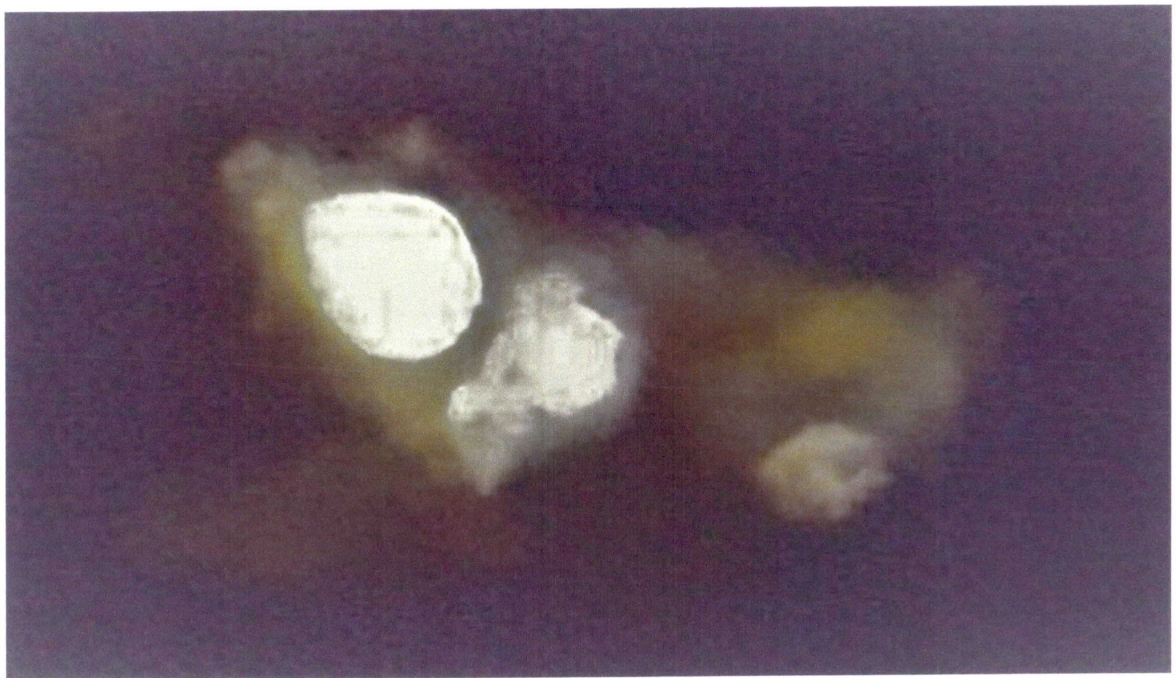

UFO Picture 1014 (Zeke Filter): April 30, 2019: 11:45:52 CST
Scott Gearen

When filters are used on the pictures, parts of the object that can't be seen with just human eyesight are exposed for viewing and analysis. The bright areas contain, what looks like a printed circuit board, commonly used in electronics to connect components to one another in a circuit.

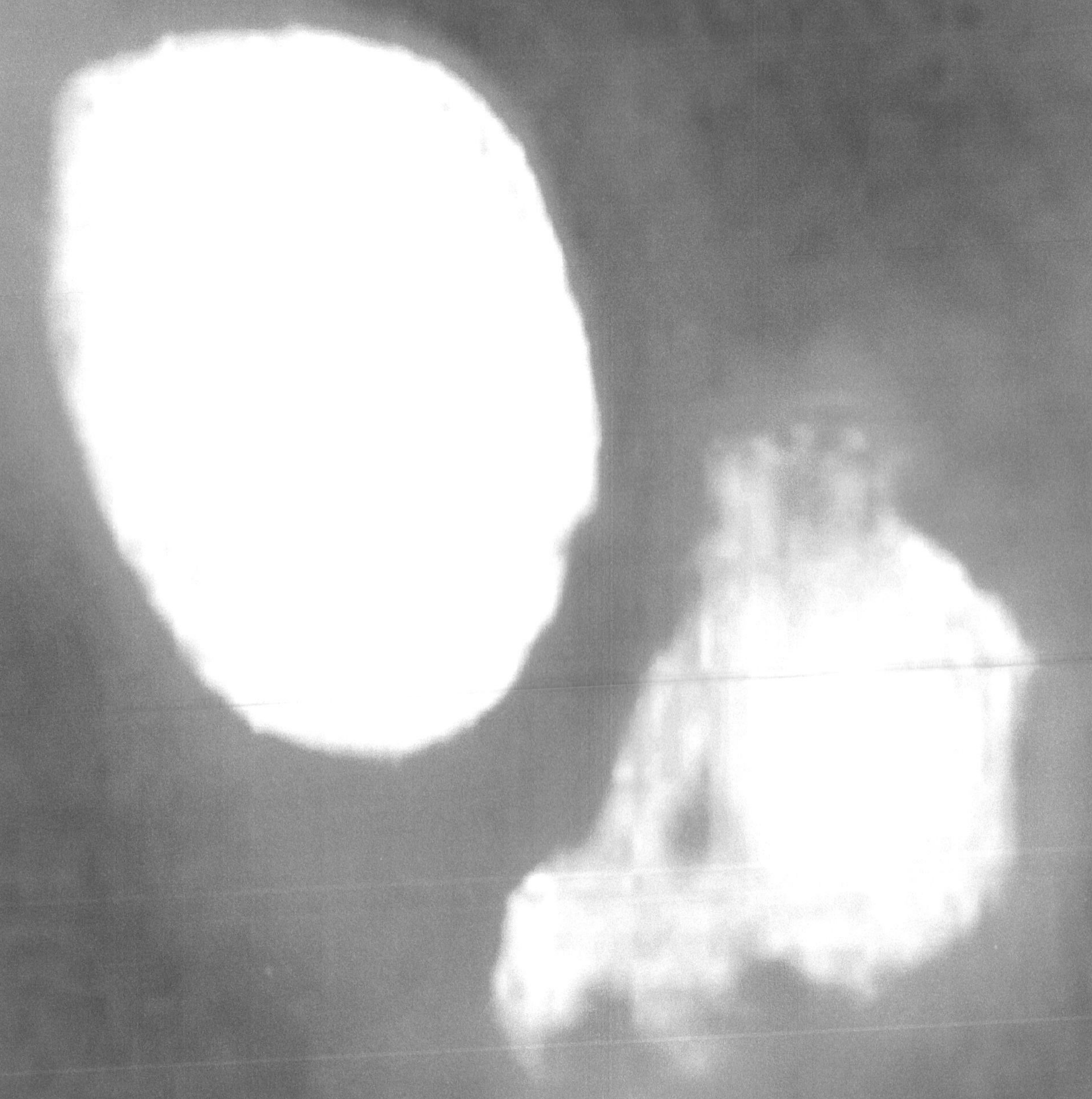

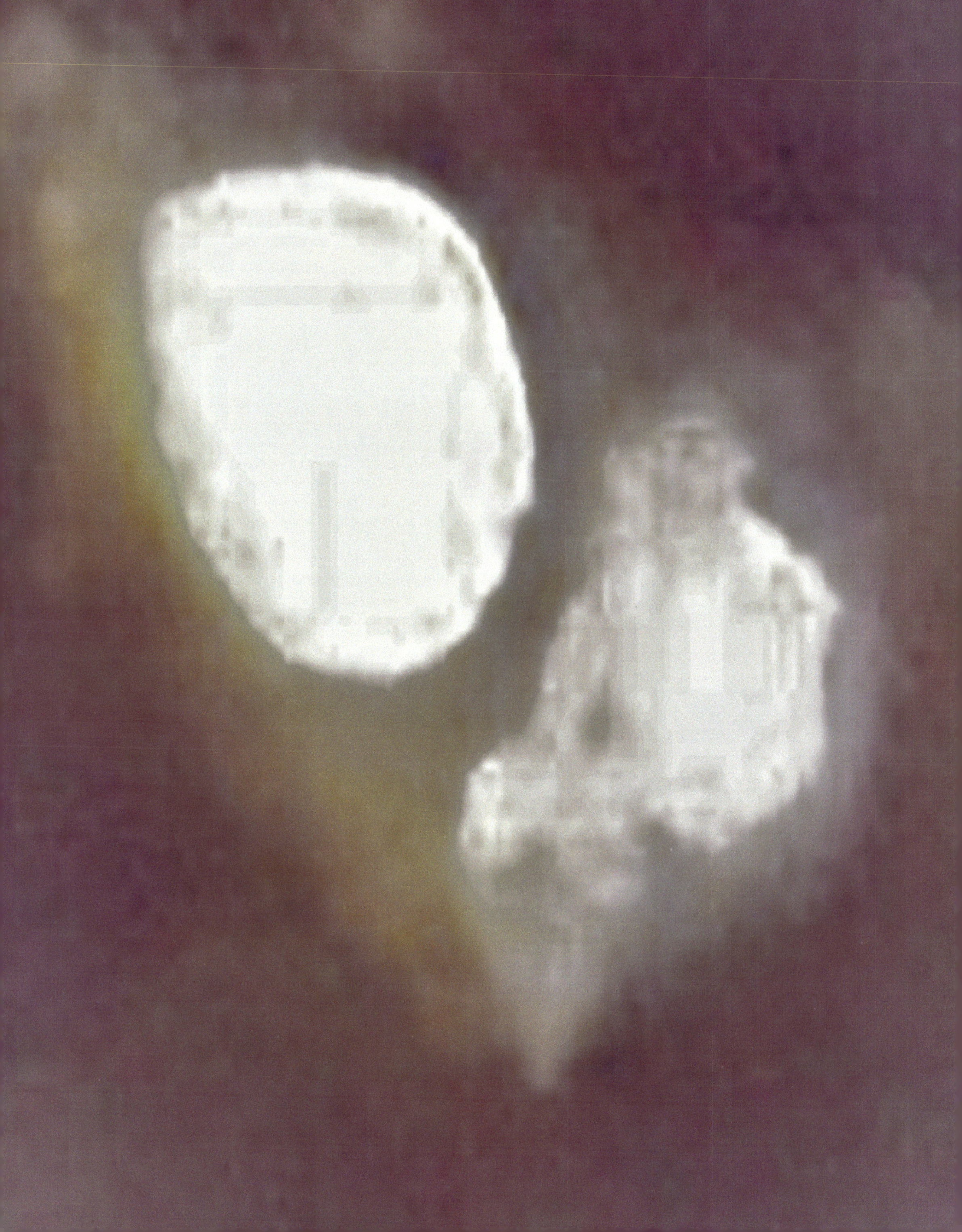

UFO Picture 1015: April 30, 2019: 11:46:00 AM CST
Scott Gearen

Eight seconds elapsed between pictures, from a distance you would not realize any changes, even if you were able to spot it above you. There are distinctive areas of the object, all continually changing shape as the object transforms. The object appears to be forming into a new shape, and a dark area is still in the center area. The pink colored object looks like it is changing at a different rate, maybe it is a mechanical part of a biological entity.

UFO Picture 1015: 30 April, 2019: 11:46:00 AM CST
Scott Gearen

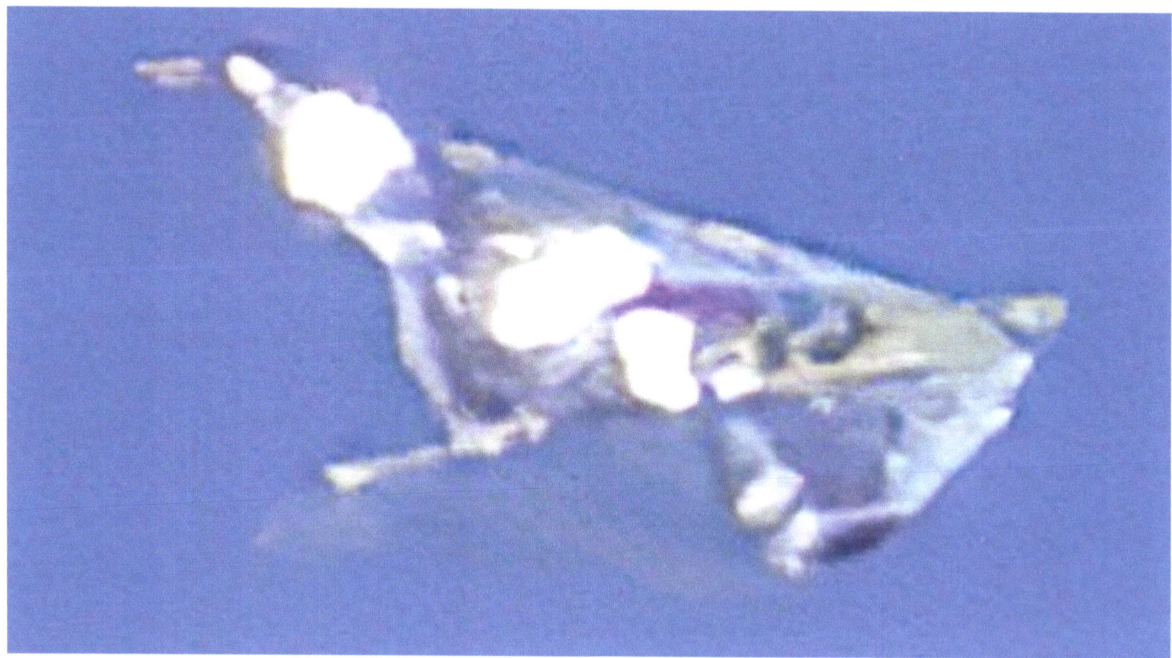

UFO Picture 1016: April 30, 2019: 11:46:22 AM CST
Scott Gearen

No longer just a blur, a new but similar shape has appeared. The object more closely resembles something from the ocean than what is expected to be in the atmosphere around us. Bright areas are present and seem to move over the body of the craft like an octopus... is it a biological species?

UFO Picture 1016: April 30, 2019: 11:46:22 AM CST
Scott Gearen

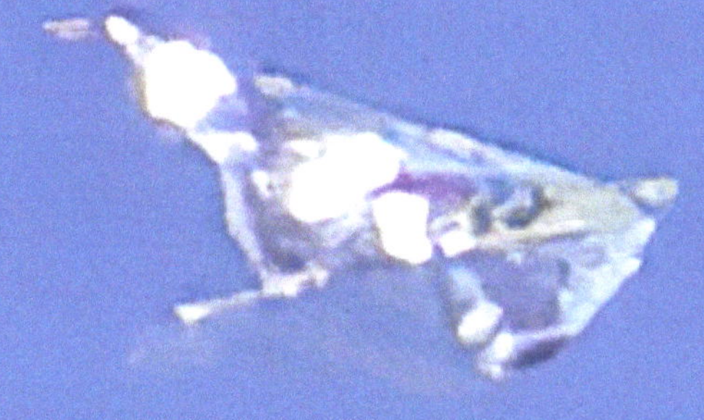

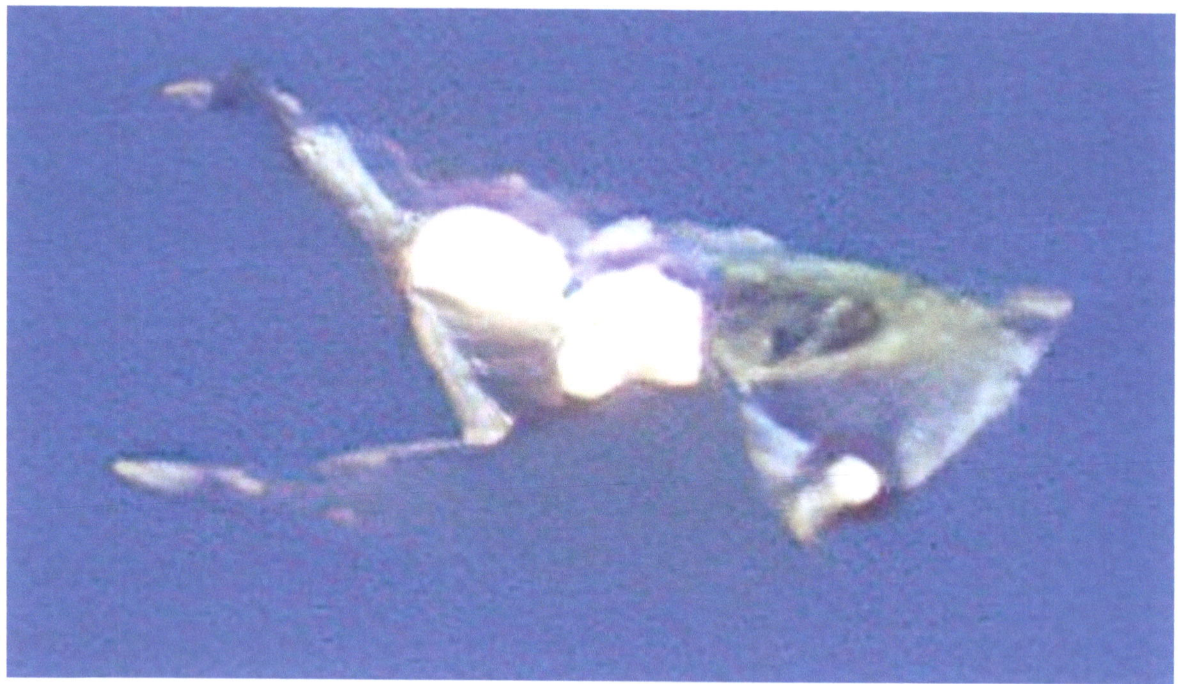

UFO Picture 1017: April 30, 2019: 11:46:24 AM CST
Scott Gearen

Each picture is different, from slight changes to a completely different shaped object. Although the object appears to be the same as the previous picture, the changes are occurring. The dark area in the center section of the object that has been present several times, it now appears to have a 3-dimensional depth to it, as if it is opening to allow access for occupants that may be inside, could there be someone or something inside... Is this a biological craft capable of dimensional travel... Is there a secret space program this is part of or is this something previously unknown.

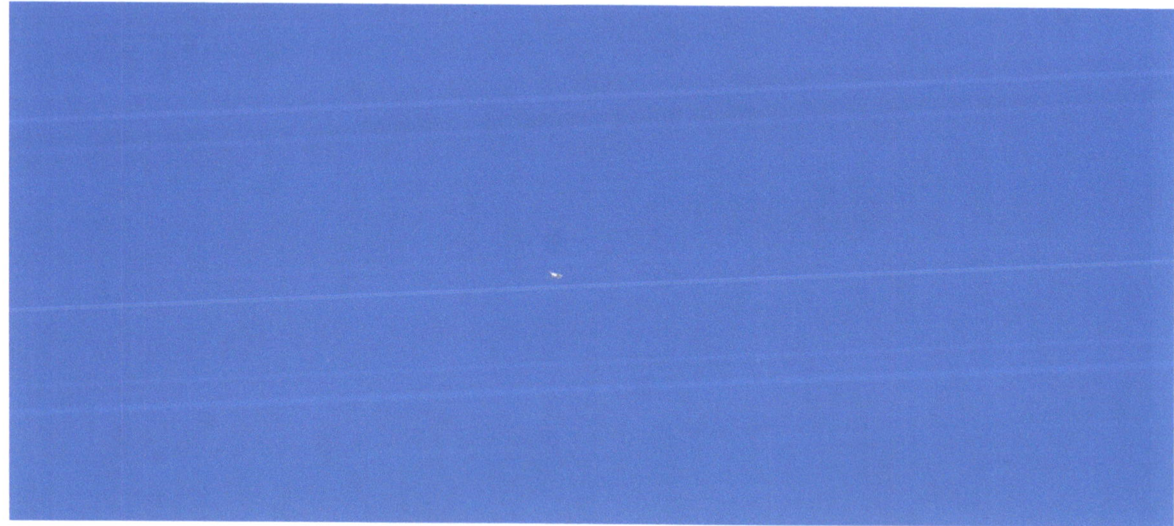

UFO Picture 1017: April 30, 2019: 11:46:24 AM CST
Scott Gearen

UFO Picture 1018: April 30, 2019: 11:46:32 AM CST
Scott Gearen

Throughout the changes there continues to be defined edges, sometimes straight and sharp, at other times wavy, as if it is being transported. Inside the edges the object appears to be a continuous flow, as if ingredients are mixed in a bowl. At times the mixture seems to appear in the shape of faces. Is this real, are there faces of an unknown race from somewhere in the universe, maybe here on earth in another dimension, or is this an illusion known as "Pareidolia"?

UFO Picture 1018, April 30 2019: 11:46:32 AM CST
Scott Gearen

UFO Picture 1018: April 30 2019: 11:46:32 AM CST
Scott Gearen

The different sections on this object seem to have 3-Dimensional depth and something that looks to me like a face wearing a Spartan helmet encased in a capsule. I used a Vanilla Filter for top picture and Zeke Filter for the bottom picture. Could this be the type of object observed and then depicted in paintings by medieval and Renaissance painters...

UFO Picture 1018: April 30 2019: 11:46:32 AM CST
Scott Gearen

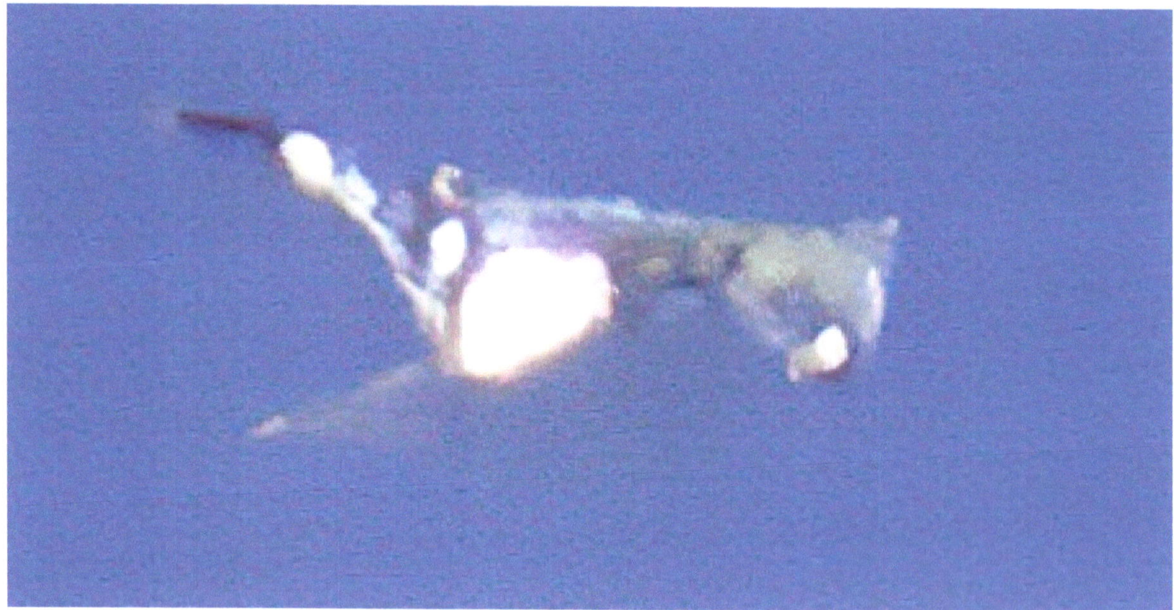

UFO Picture 1019: April 30, 2019: 11:46:42 AM CST
Scott Gearen

There seems to be two distinct phases during this series of pictures. A very active phase and somewhat of a resting phase. In the resting phase, the object has a very distinct shape observable for a few seconds, then the bright areas increase both intensity and size and the object goes into an active phase, it increases until is appears as a single object, then disappears and suddenly returns to view as a new and different shaped object. Due to the increased size and intensity of the bright areas the object is going into the active phase of transformation.

UFO Picture 1019: April 30, 2019: 11:46:42 AM CST
Scott Gearen

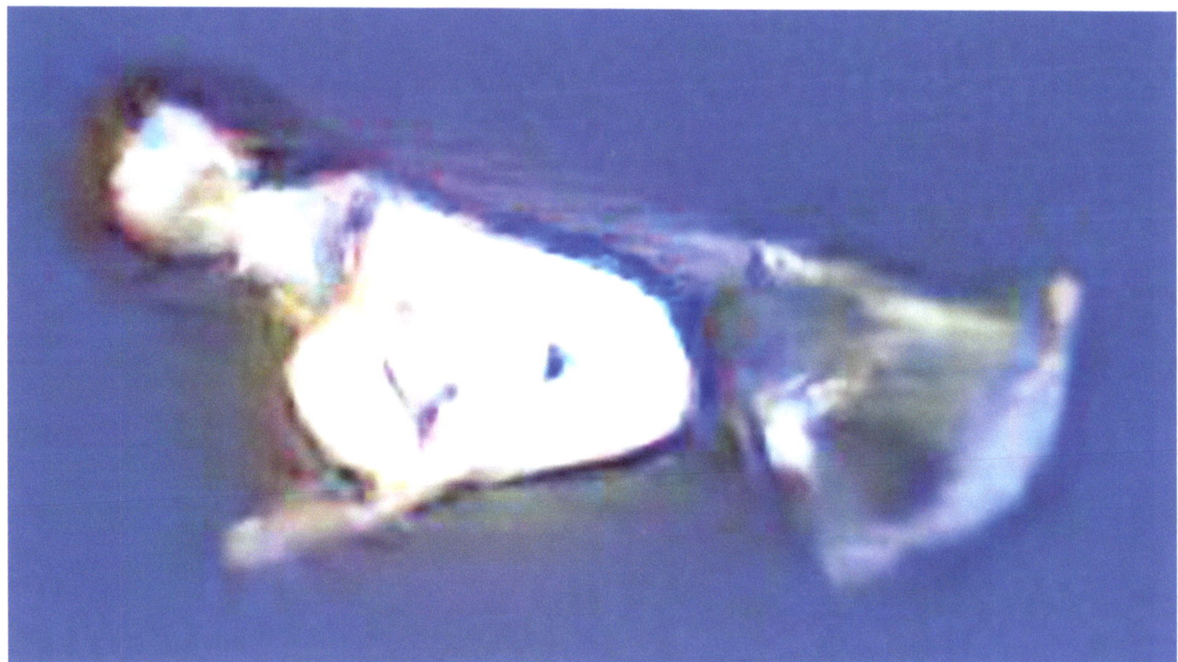

UFO Picture 1020: April 30, 2019: 11:46:52 AM CST
Scott Gearen

The bright area seems to enlarge to a point where the entire object becomes a single large spherical ball of light. It seems to be a transformation that is beyond human comprehension as to what it is, what it is doing, how it is doing whatever it is doing, or where it came from. Metamorphosis is taught in science classes as the process of transformation of insects, animals or aquatic species from hatching or birth into an adult. The metamorphosis of this object is a phenomenon that may not ever have been previously observed, except by governments and ET.

UFO Picture 1020: April 30, 2019: 11:46:52 AM CST
Scott Gearen

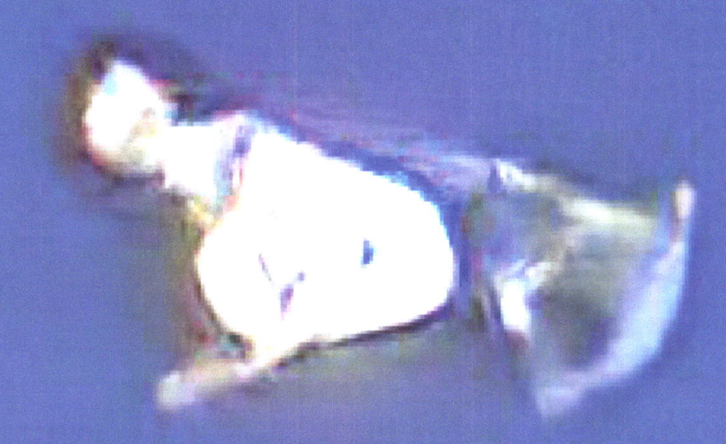

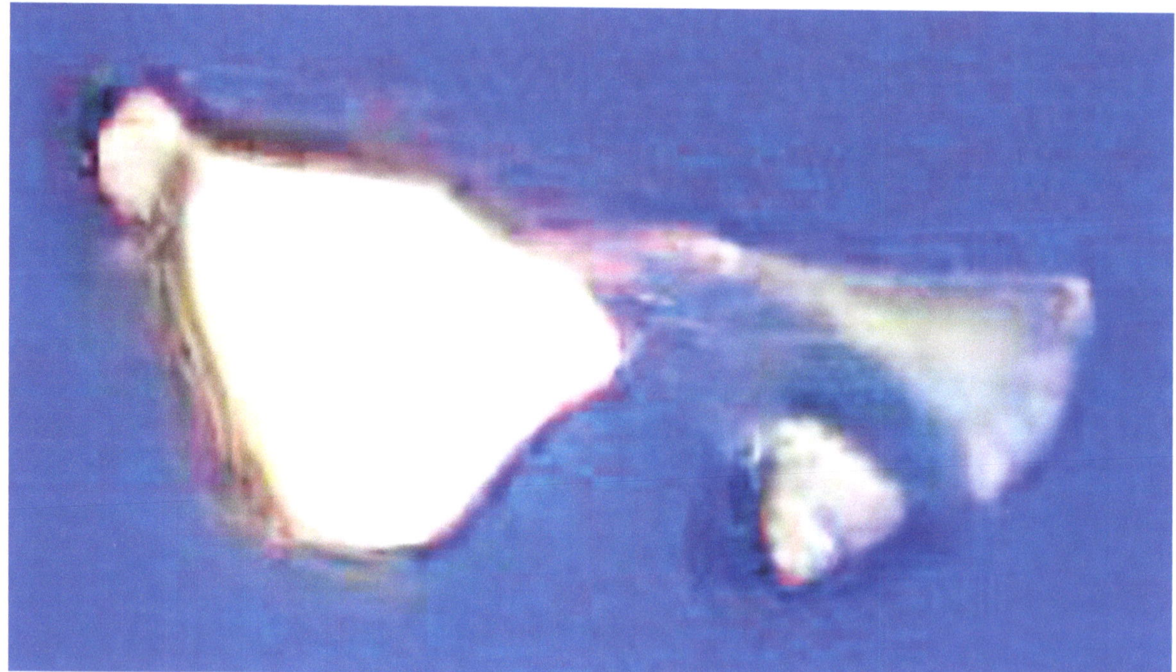

UFO Picture 1021: April 30, 2019: 11:46:56 AM CST
Scott Gearen

As the object is consumed by the extreme bright areas it appears the outer edges of the object are beginning to compress inward, as if being pulled into the bright area. Simultaneously a distinctive circular area in the center section of the object is forming, with a spiral appearance to it. What is caught on camera in this picture, could be the prelude to the many pictures that capture the bright spheres, as well as the rings that appear to be spinning at a high rate of speed.

UFO Picture 1021: April 30, 2019: 11:46:56 AM CST
Scott Gearen

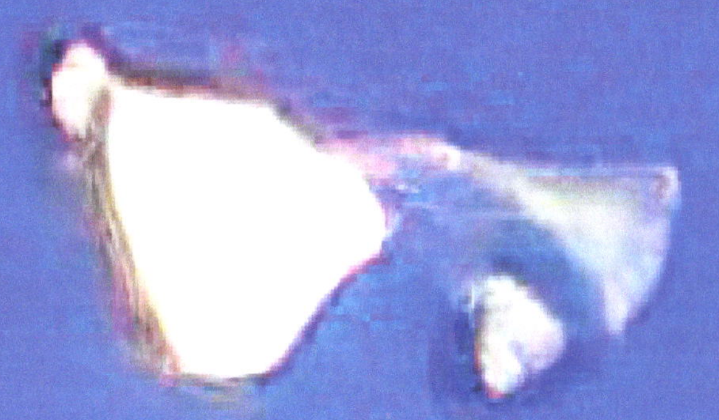

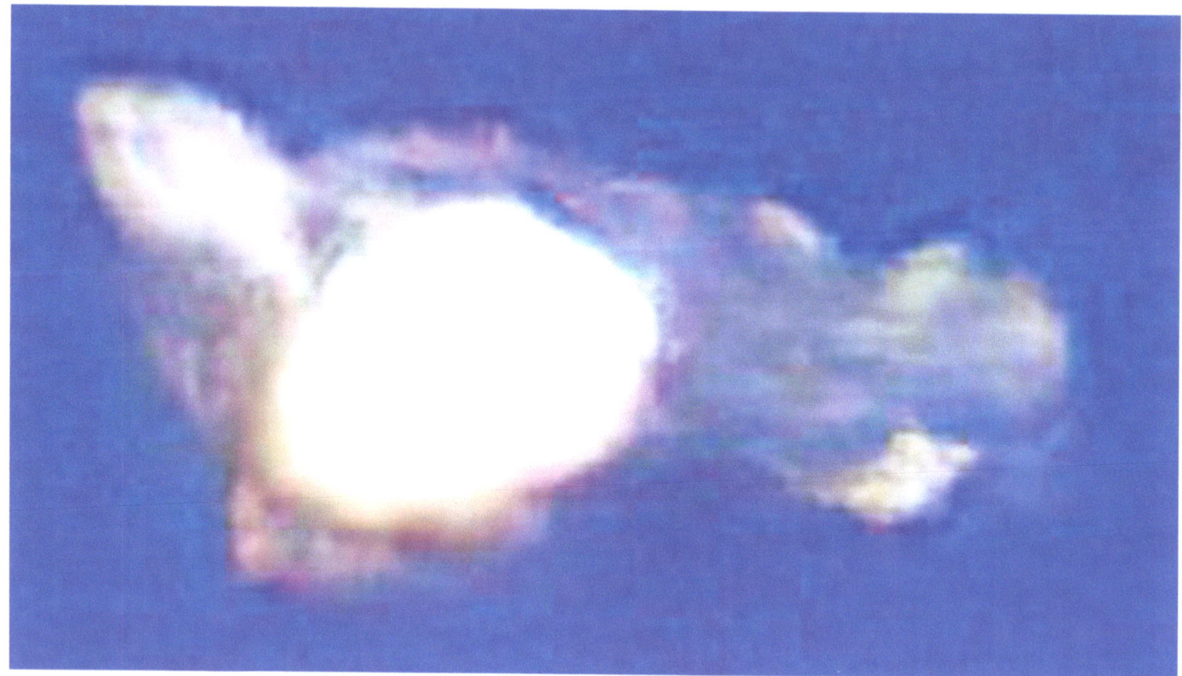

UFO Picture 1022: April 30, 2019: 11:47:06 AM CST
Scott Gearen

Is this an amazing display of a natural occurrence that has been going on for millenniums, or is this something that has been created, and if it is created, who created it? Why was I able to see it and by fortune to have a camera with me, for without it no one else would have seen it, nor believed me. The contrail to the left is in focus, the object is in focus, the movement and blending of the object make it appear to be out of focus.

UFO Picture 1022: April 30, 2019: 11:47:06 AM CST
Scott Gearen

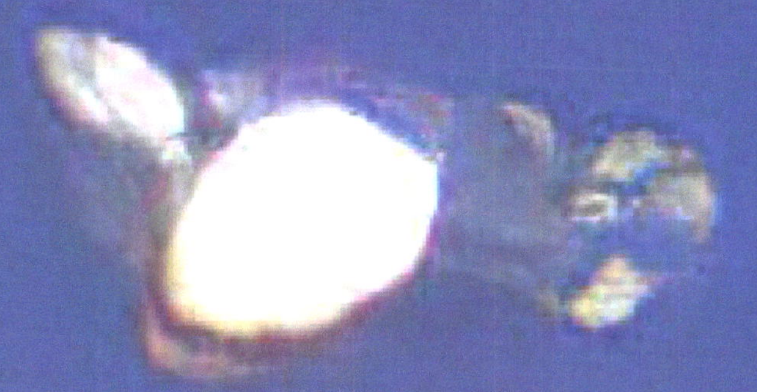

UFO Picture 1023: April 30, 2019: 11:47:18 AM CST
Scott Gearen

Stop here for a few seconds, turn the pages back to picture 1019 and observe a typical cycle in this entire series. From picture 1019 to picture 1023 there has only been an elapsed time of 36 seconds for this object to go through a complete transformation from one shape to the next.

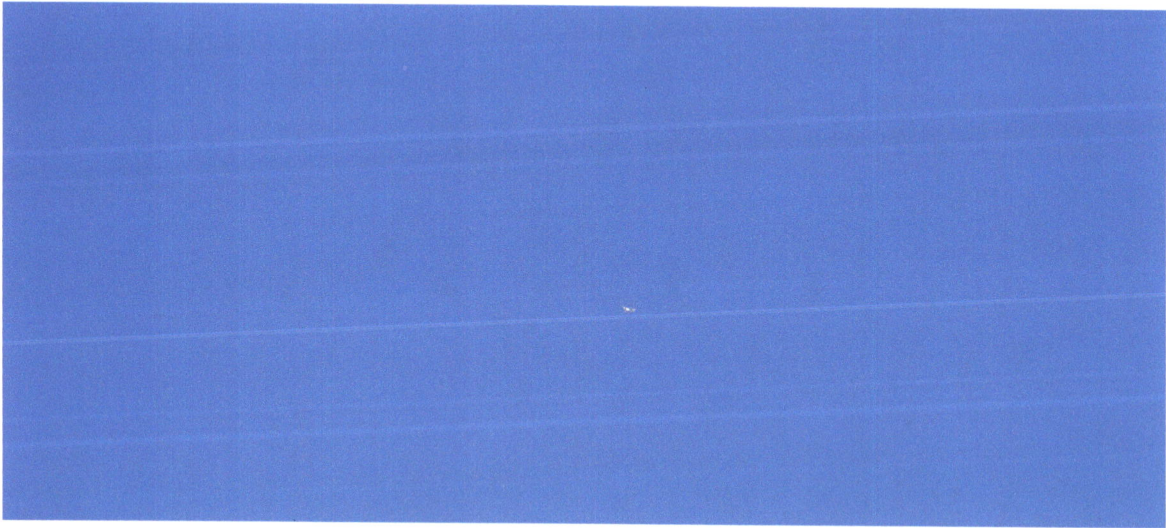

UFO Picture 1023: April 30, 2019: 11:47:18 AM CST
Scott Gearen

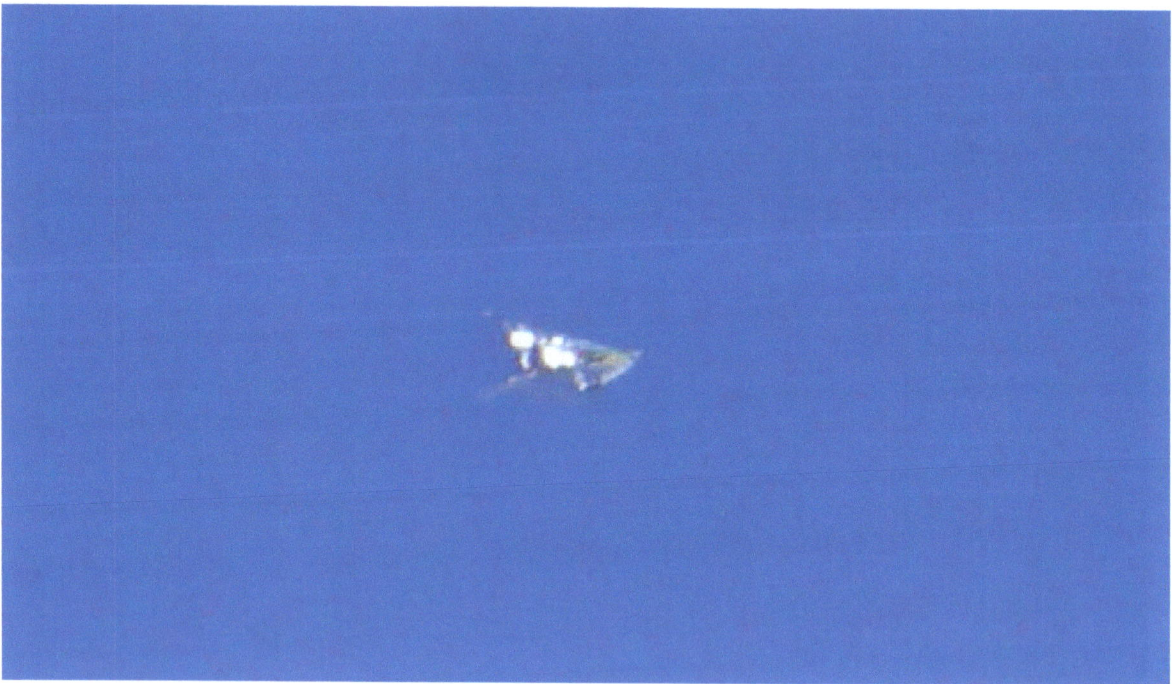

UFO Picture 1024: April 30, 2019, 11:47:30 AM CST
Scott Gearen

The object seems to be resting, what I refer to as the "passive" phase of these transformations. There are minimal changes from the previous pictures, the most obvious changes are the bright areas, these areas are always in an "active" phase as they increase size, intensity, and movement. Are the bright areas a source of power? The bright rectangle shape looks like a generator.

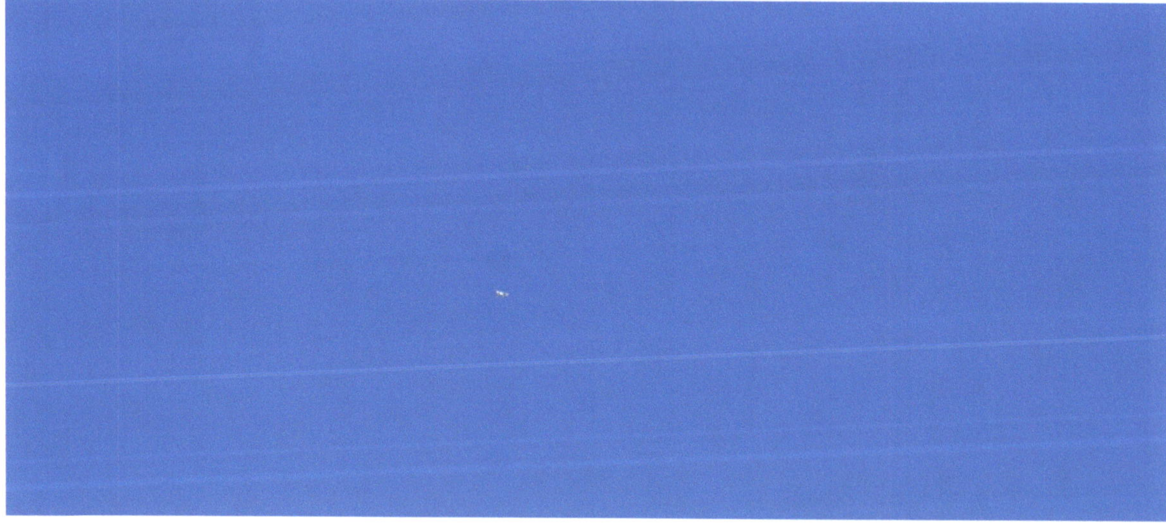

UFO Picture 1024: April 30, 2019, 11:47:30 AM CST
Scott Gearen

UFO Picture 1025: April 30, 2019: 11:47:42 AM CST
Scott Gearen

An active phase of transformation is beginning, the bright areas are increasing in size and intensity, soon the object will be engulfed in these phenomena, and a new shaped object will appear. The large black area, and a protrusion on the opposite side, which I think are "eyes" continue to appear in the same area. The underside of the wing is the color of the sky... is this to camouflage itself... I think it is intentional to hide itself from below, a clear sign of intelligence.

UFO Picture 1025: April 30, 2019: 11:47:42 AM CST
Scott Gearen

UFO Picture 1026: April 30, 2019: 11:47:52 AM CST
Scott Gearen

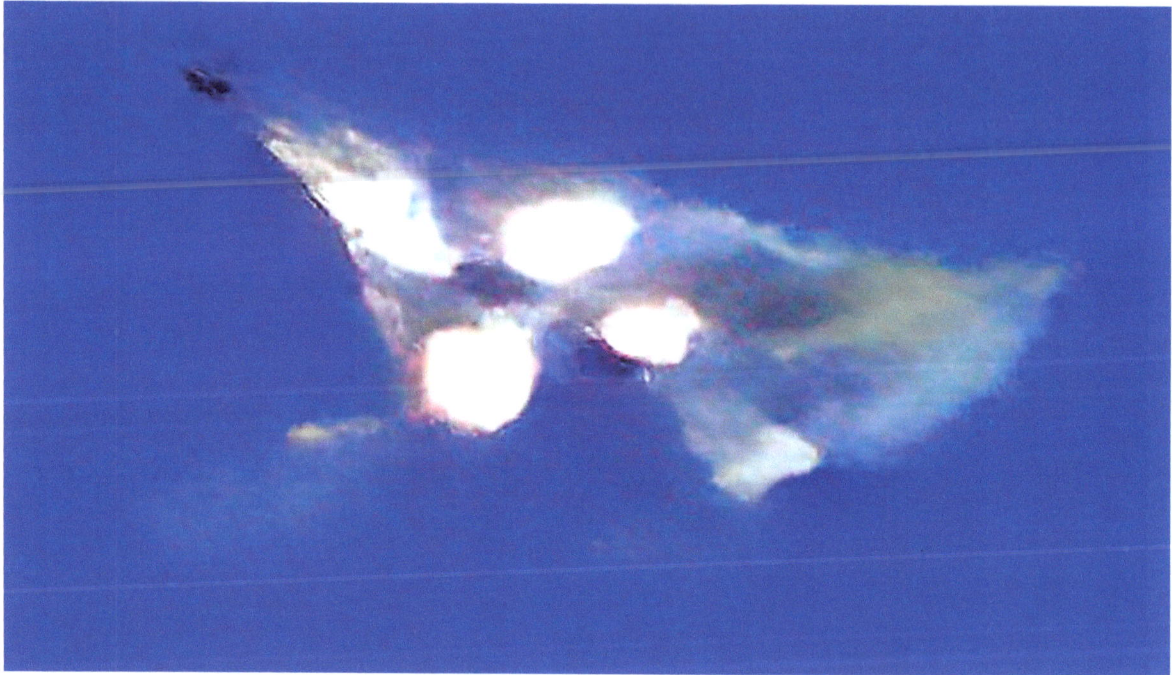

UFO Picture 1026: April 30, 2019: 11:47:52 AM CST
Scott Gearen

The entire object is beginning to blend together, the center area which may be a capsule for occupants remains intact. The bright areas have moved into separate and distinct areas, this movement can be seen in a video that was taken during this sighting. The different colors imply to me there is a chemical reaction occurring with different gases that may be within the object.

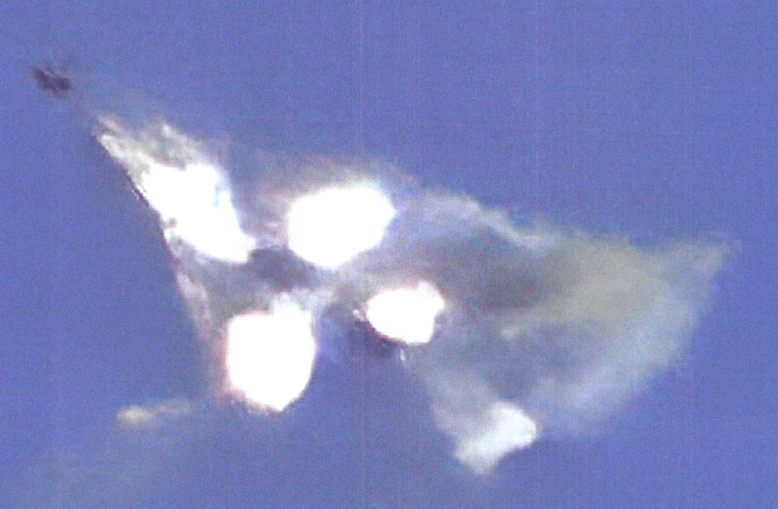

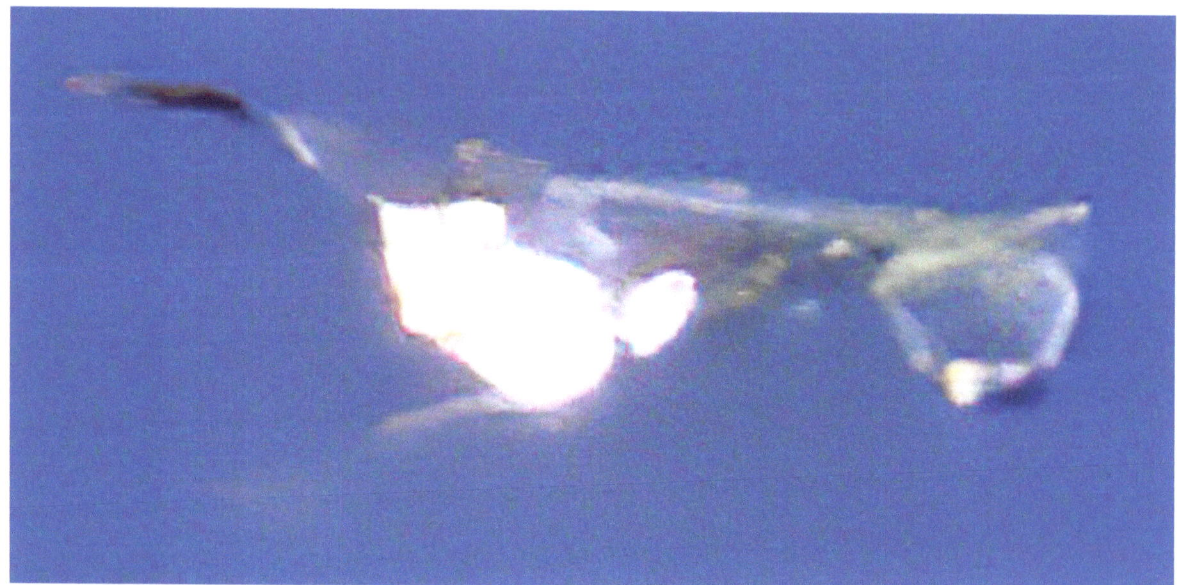

UFO Picture 1027: April 30, 2019: 11:48:02 AM CST
Scott Gearen

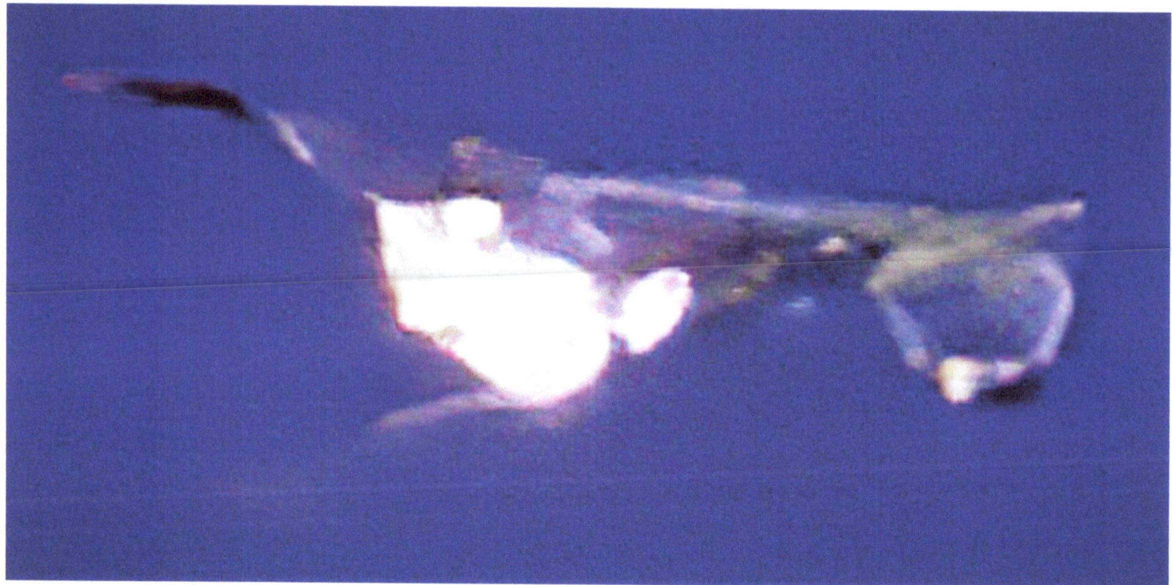

UFO Picture 1027: (Rouge Filter): April 30, 2019: 11:48:02 AM CST
Scott Gearen

The speed at which the bright areas move across the outer skin of the craft, which is similar to an octopus changing colors to match its background, is too fast to be seen with eyesight alone, without the camera stopping the action this looked like a flashing bright light in the air. Everything at this point in time is purely speculation as to what this is. The object doesn't need forward momentum to maintain altitude, nor is there a noticeable power source keeping it aloft. Whatever the object is, it has the ability to defy gravity. The dark area in the center once again has a bit of depth to it, almost as if I am able to peer inside, as well as the ability for an occupant to peer out …

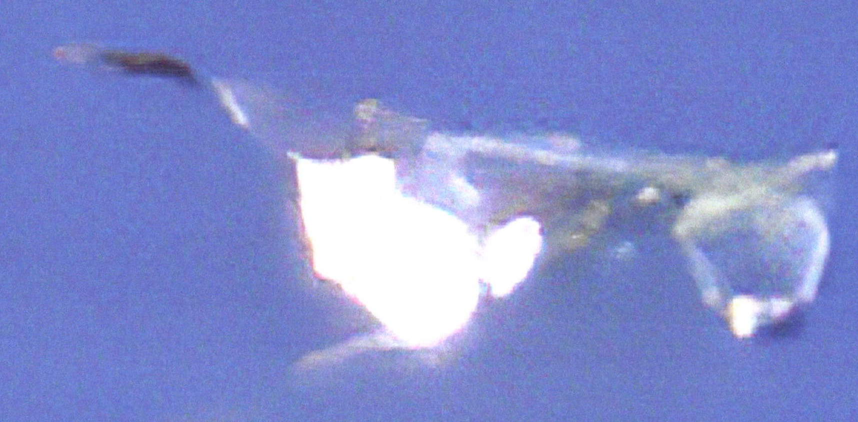

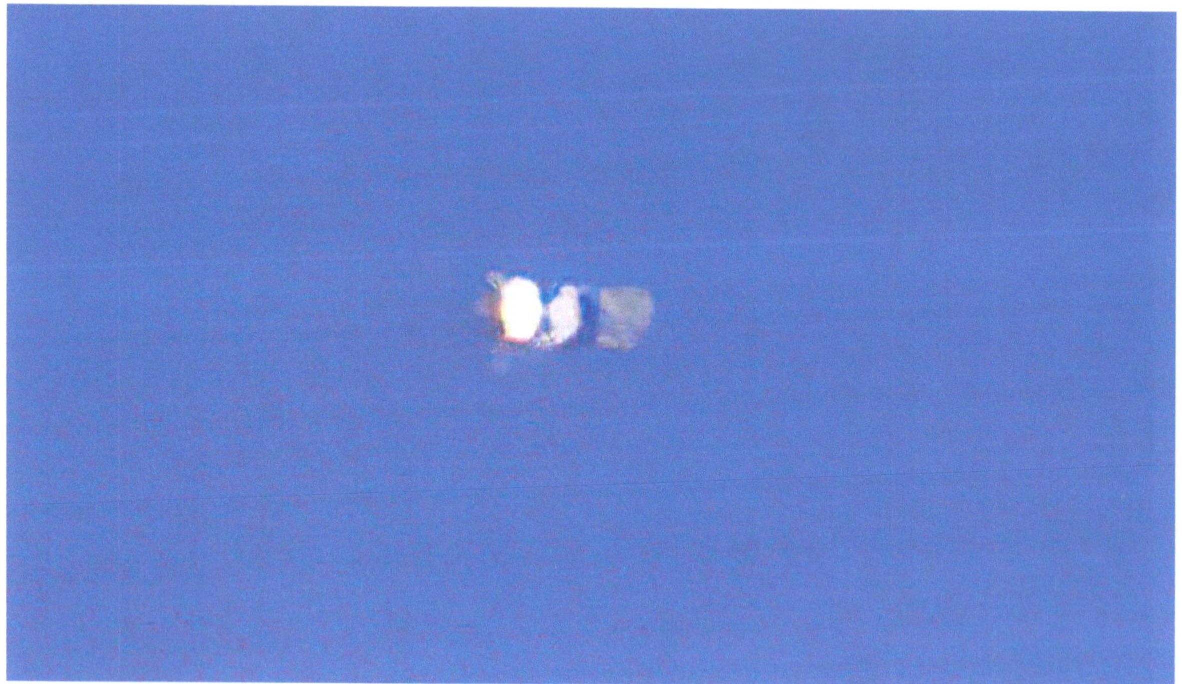

UFO Picture 1028: April 30, 2019: 11:48:10 AM CST
Scott Gearen

UFO Picture 1028: April 30, 2019: 11:48:10 AM CST
Scott Gearen

When I saw this my first thought was the object has split itself into two objects, which is often reported during sightings. Later as I thought about this I wondered how or why it would do this, and the thought occurred to me; "energy". This is a nuclear reaction, specifically nuclear fission. Hundreds of nuclear reactors around the world do this every day producing energy for the world.

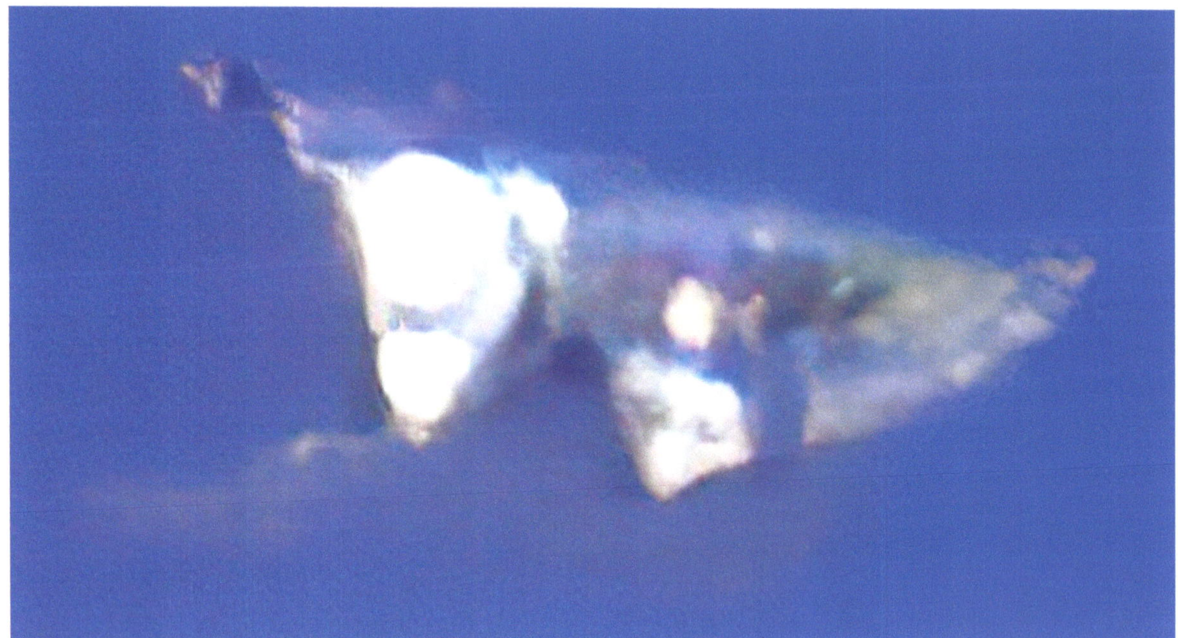

UFO Picture 1029: April 30, 2019: 11:49:48 AM CST
Scott Gearen

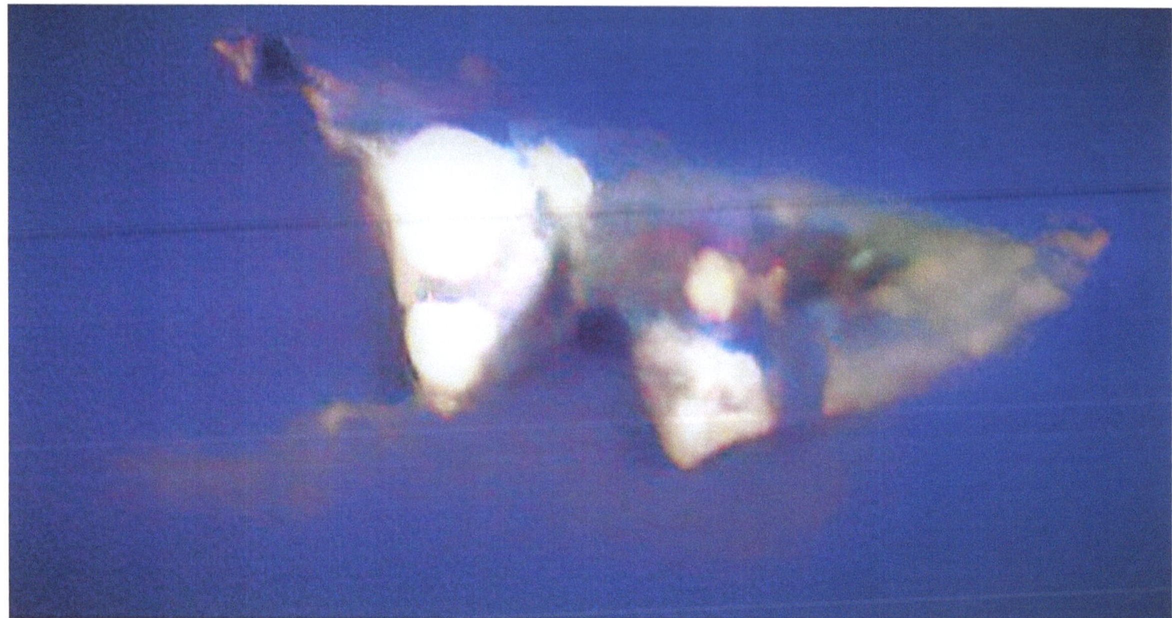

UFO Picture 1029 (Napa Filter): April 30, 2019: 11:49:48 AM CST
Scott Gearen

This picture was taken over one minute since the last picture, based on the metadata from previous pictures this object has been continuously changing shape within seconds, and based on the next picture it appears this object is beginning an active phase and will be consumed within a ball of light within seconds. In nuclear fission energy is released as heat or light, this object is continually exhibiting both heat and light. It is possible this object is capable of producing its own energy source and is the reason for the continuous changes that appear as a very bright light.

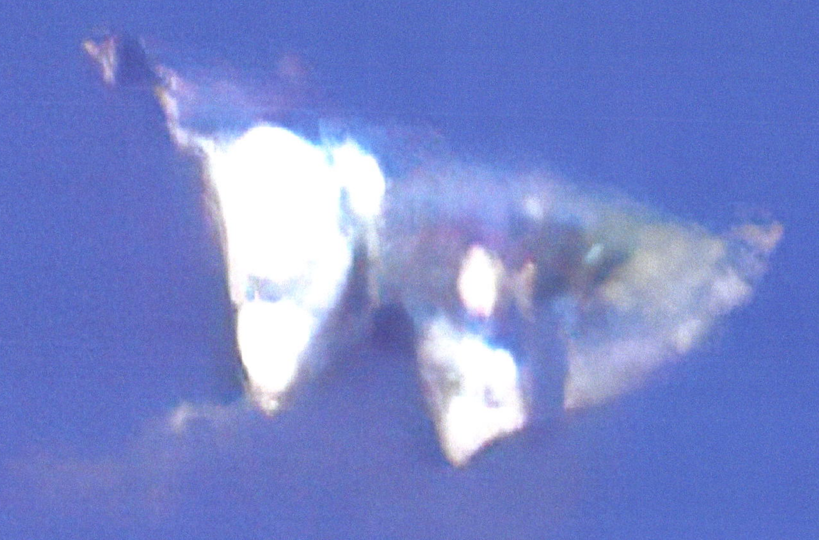

UFO Picture 1030: April 30, 2019: 11:49:58 AM CST
Scott Gearen

A spinning nucleus produces heat, energy, and creates a magnetic field producing... "anti-gravity". Is that a printed circuit board within the sphere surrounded by an energy field.

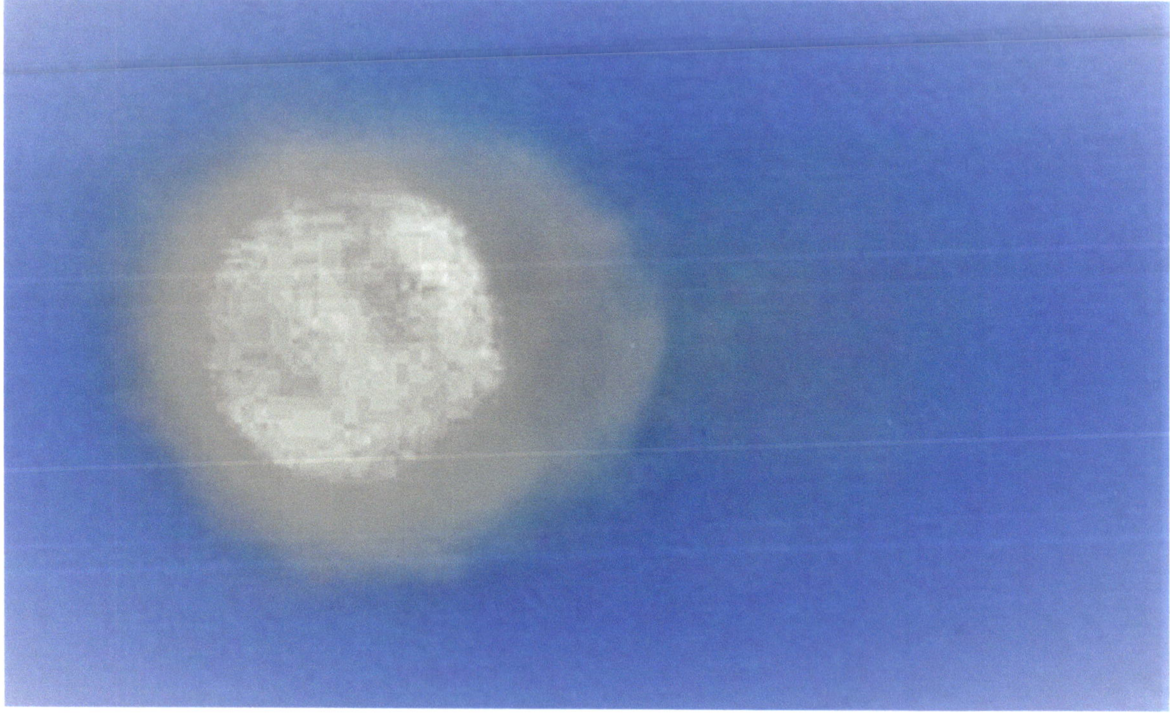

UFO Picture 1030 (Icarus Filter): April 30, 2019: 11:49:58 AM CST
Scott Gearen

UFO Picture 1031 (Denim Filter): April 30, 2019: 11:50:18 AM CST
Scott Gearen

Pictures are enlarged and have Denim filter applied. Seeing an object fade into and out of view (Top) reminds me of the Philadelphia experiment, where it is alleged that in 1943 a U.S. Navy destroyer was made invisible and teleported from Philadelphia Pennsylvania to Norfolk Virginia. Both pictures appear to show a type of circuit board in the bright area that is similar to what is found in electronic equipment, such as computers.

UFO Picture 1031 (Denim Filter): April 30, 2019: 11:50:18 AM CST
Scott Gearen

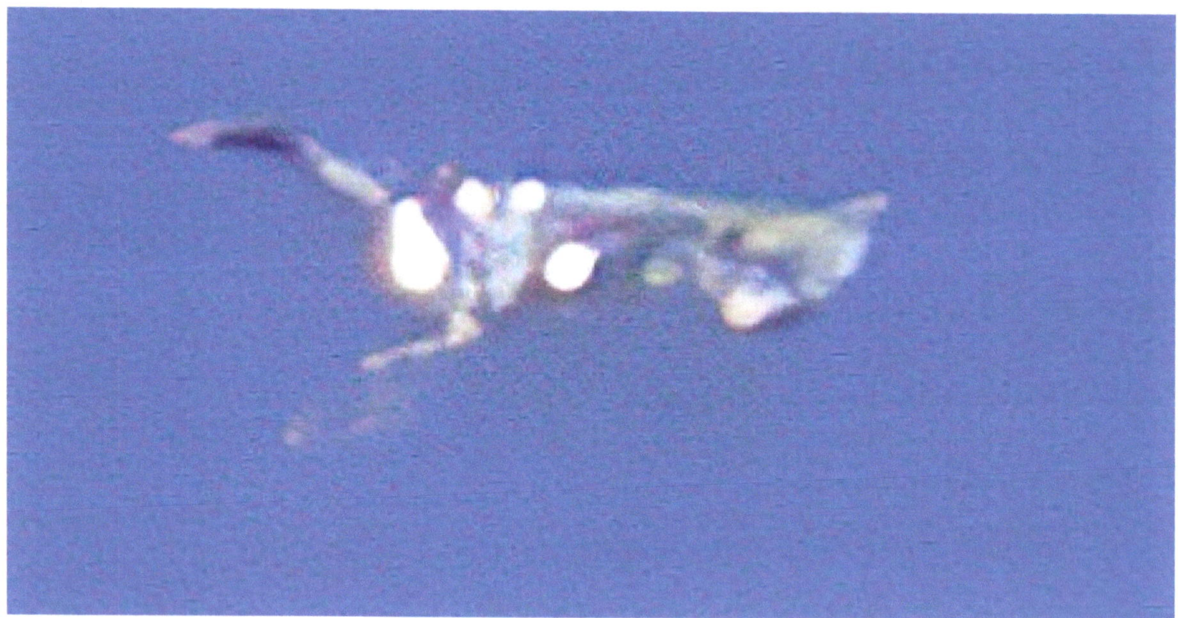

UFO Picture 1032: April 30, 2019: 11:50:28 AM CST
Scott Gearen

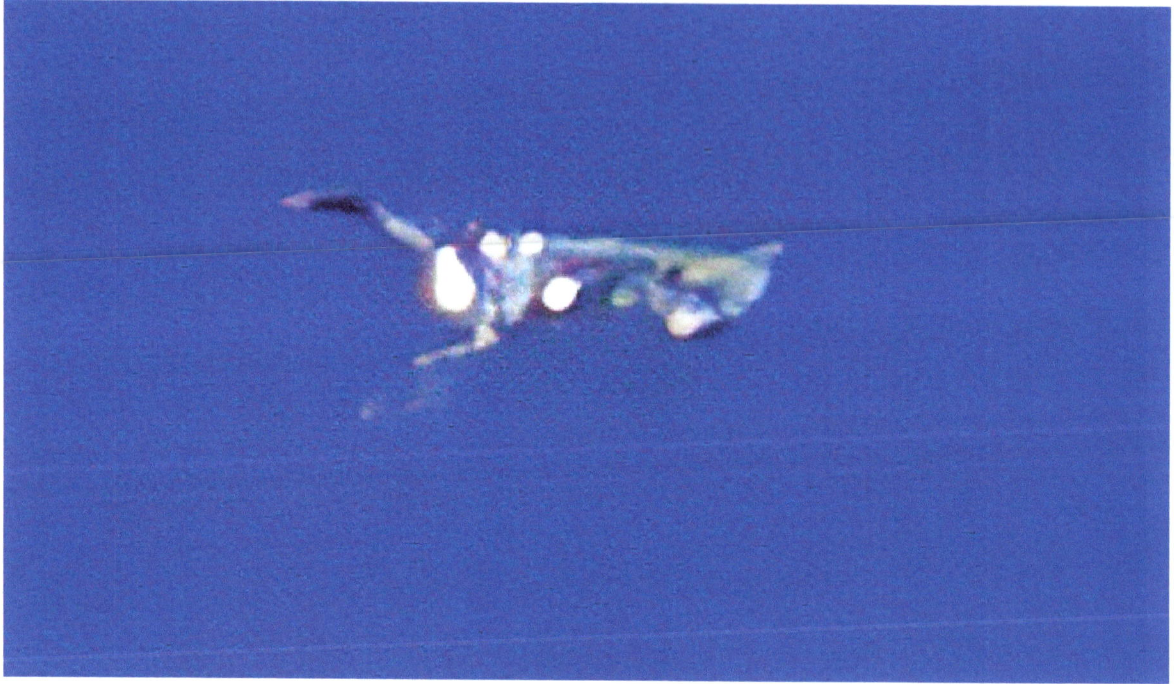

UFO Picture 1032 (Enhanced Exposure): April 30, 2019: 11:50:28 AM CST
Scott Gearen

How could this object get any stranger… in just ten seconds it has transformed from an object that on one side is half blinding light and the other half can barely be seen as it fades from view, to something that appears to be a large creature with wings, black eyes, and a vertical tail floating miles above in the air. Could this be a crypto-creature living here on earth, or is this an Extraterrestrial Biological Being (EBE) from somewhere in the universe, or is it a craft for space travel?

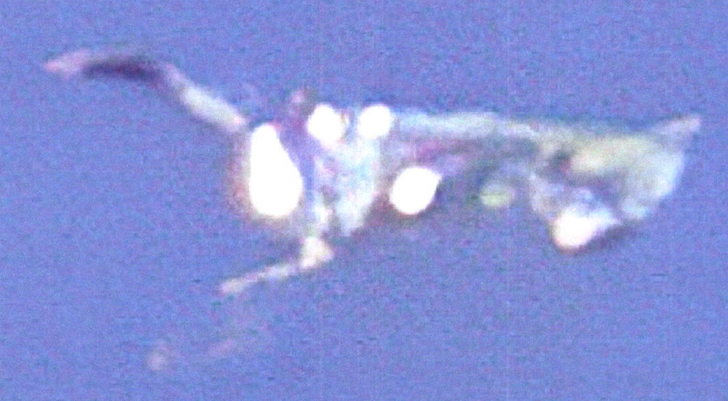

UFO Picture 1033: April 30, 2019: 11:50:36 AM CST
Scott Gearen

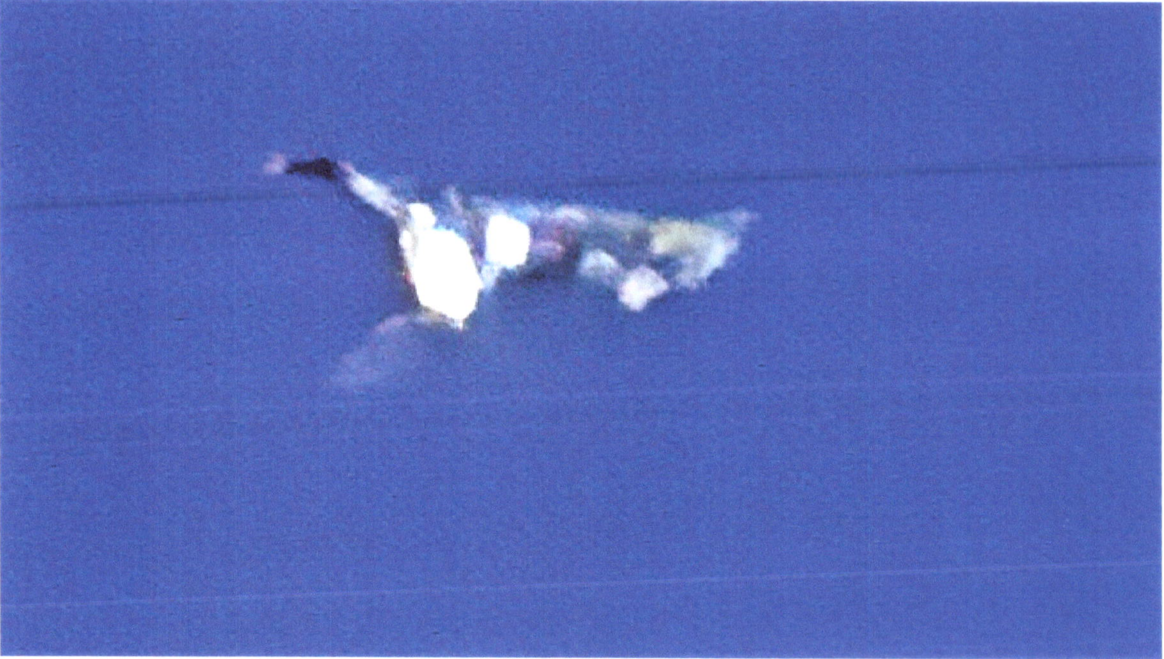

UFO Picture 1033 (Enhanced Lighting): April 30, 2019: 11:50:36 AM CST
Scott Gearen

The object looks similar to the previous picture, the subtle changes, although they occur quicky, are easily missed. The bright areas are noticed immediately, but the real change is in the body of the craft, the object seems almost as if it is becoming liquified, the colors are more vibrant, and seem to be blending together, which means an active cycle is coming quickly.

UFO Picture 1034 (Zeke Filter): April 30, 2019: 11:50:44 AM CST
Scott Gearen

This picture (1034) was taken eight seconds after picture 1033. During my observation of this phenomena there were moments when the object could not be seen with regular eyesight and required polarized sunglasses to see the object. My camera lens was not able to pick it up and only through use of filters was the object able to be located as a barley visible white sphere.

UFO Picture 1034: April 30, 2019: 11:50:44 AM CST
Scott Gearen

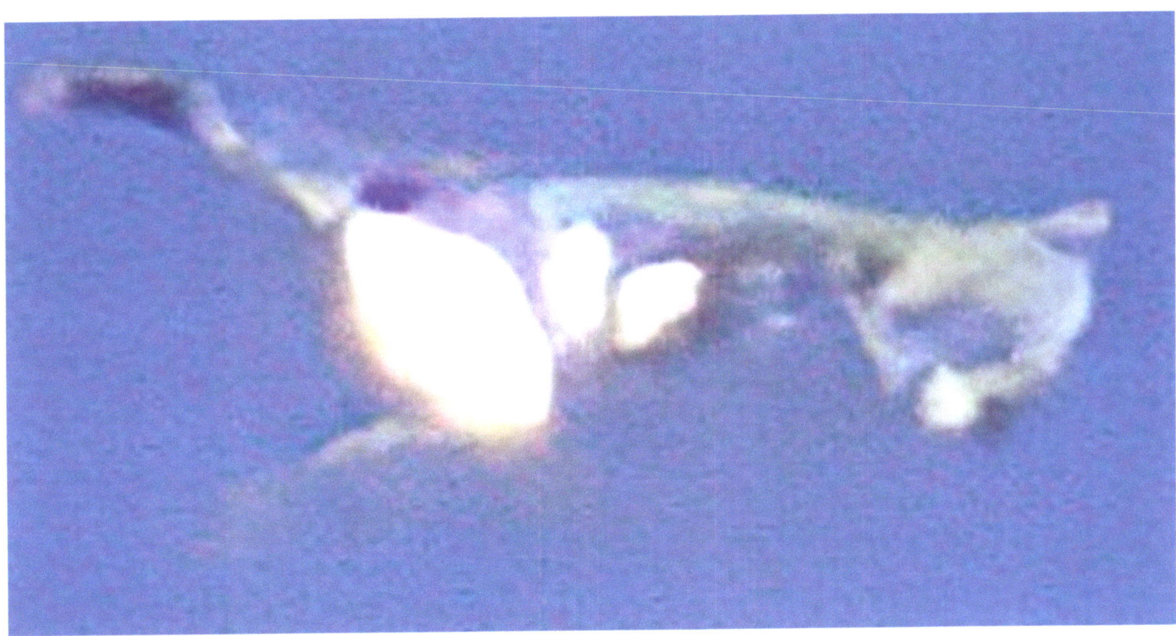

UFO Picture 1035: April 30, 2019: 11:50:56 AM CST
Scott Gearen

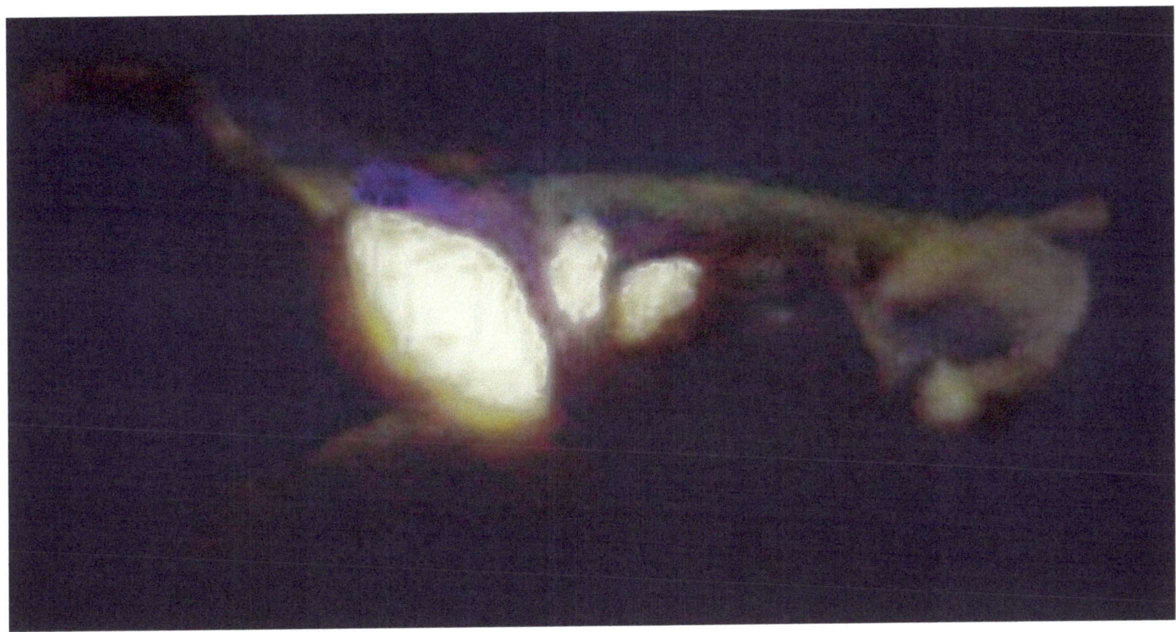

UFO Picture 1035 (Sunscreen Filter): April 30, 2019: 11:50:56 AM CST
Scott Gearen

Twelve minutes later, based on the pattern of this object there must have been several cycles not observed. This is very similar to others after a transformative cycle, several bright, intense areas, which appear throughout the cycle and usually at the end of the resting phase, right before it disappears into a sphere. The area in the middle, generally an oblong shape with a darker color, the "capsule", again appears to have depth to it. Several filters were applied to this during analysis and it appears to be occupied...

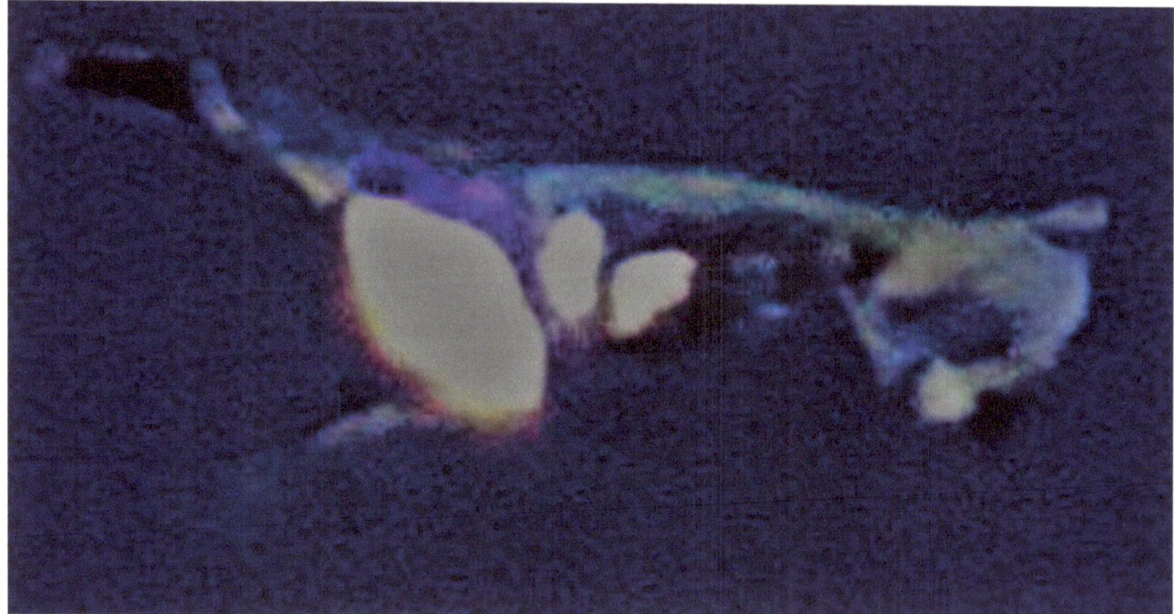

UFO picture 1035 (Sauna Filter): April 30, 2019: 11:50:56 AM CST
Scott Gearen

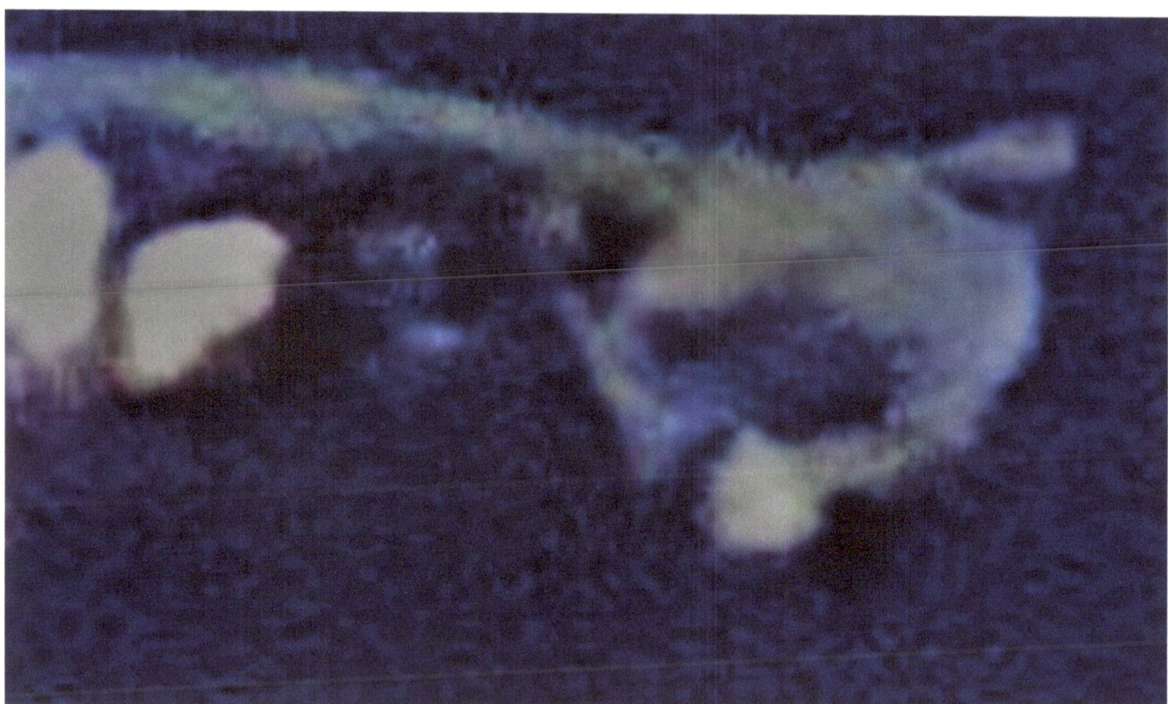

UFO Picture 1035 (Sauna Filter): April 30, 2019: 11:50:56 AM CST
Scott Gearen

This particular filter was able to provide more contrast than other filters around the "capsule" area, and appears to show two occupants inside looking out. This could be nothing more than pareidolia, due to the lighter areas and the darker areas forming figures that my mind is used to seeing. However, my mind is not used to seeing an unidentified object in the air above me, and "all options are on the table".

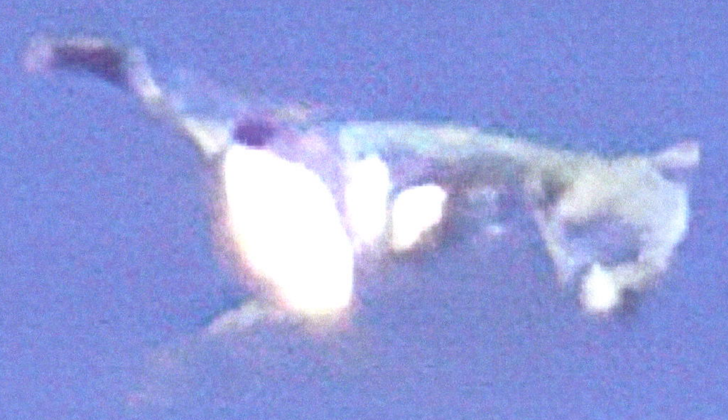

UFO Picture 1036: April 30, 2019: 11:51:18 AM CST
Scott Gearen

This picture (1036) appears to be coming out of an active phase. It was captured on camera between two pictures that both show a well-defined object in the space of approximately 30 seconds. It doesn't appear to be "cloaking" itself, rather to me, it appears this object has faded from view and is reappearing. Almost as if it is being transported particle by particle in and out of more than one location, or dimension.

UFO Picture 1036: April 30, 2019: 11:51:18 AM CST
Scott Gearen

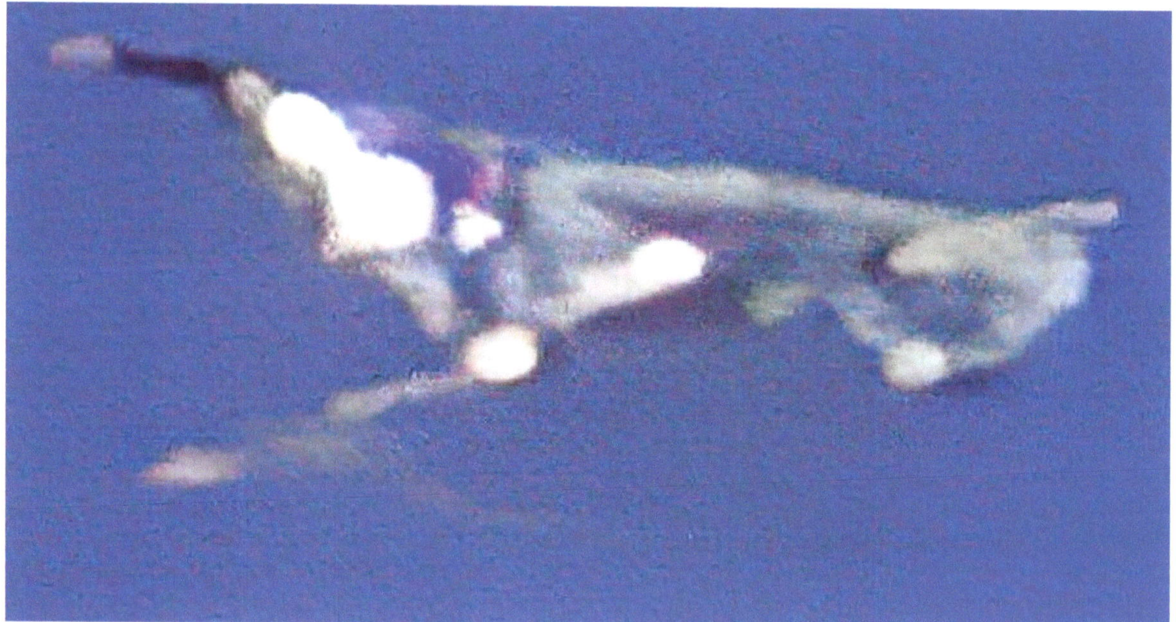

UFO Picture 1037: April 30, 2019: 11:51:26 AM CST
Scott Gearen

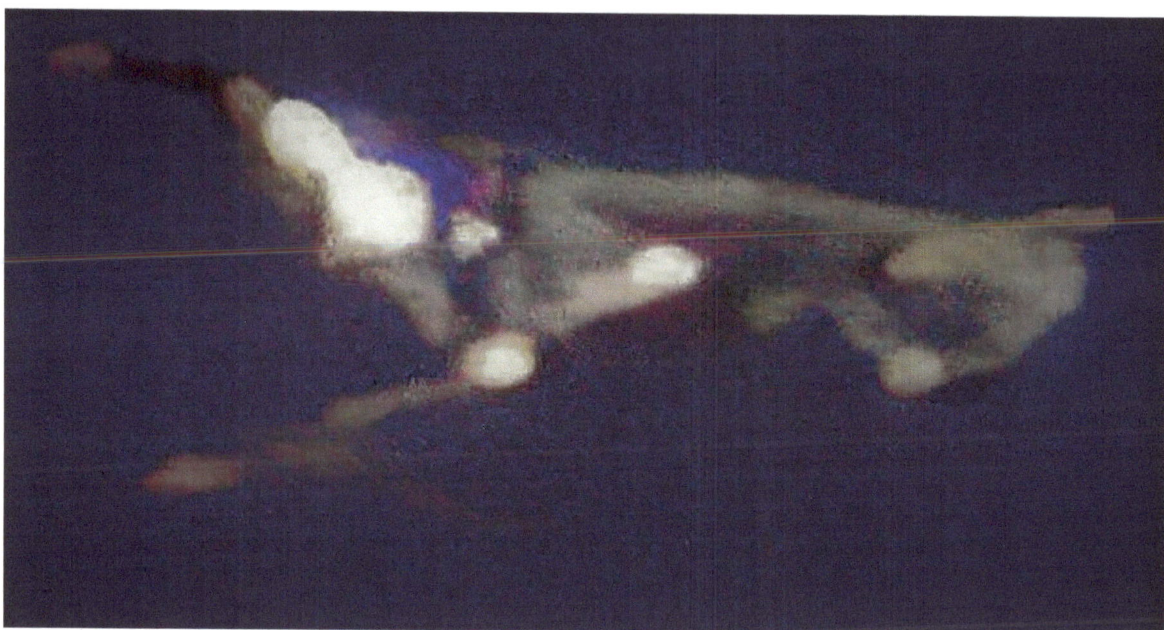

UFO Picture 1037 (Sunscreen Filter): April 30, 2019: 11:51:26 AM CST
Scott Gearen

Now, in just eight seconds, it is fully visible, the bright areas are generally smaller and more scattered right after it appears, maybe resting phase is not correct, just a different phase since it is continually changing. The dark capsule area in the middle appears closed and the colors around it appear to be moving, blending together. The bottom of the object is blue and the top is brightly colored with several of the intense spheres.

UFO Picture 1038: April 30, 2019: 11:51:36 AM CST
Scott Gearen

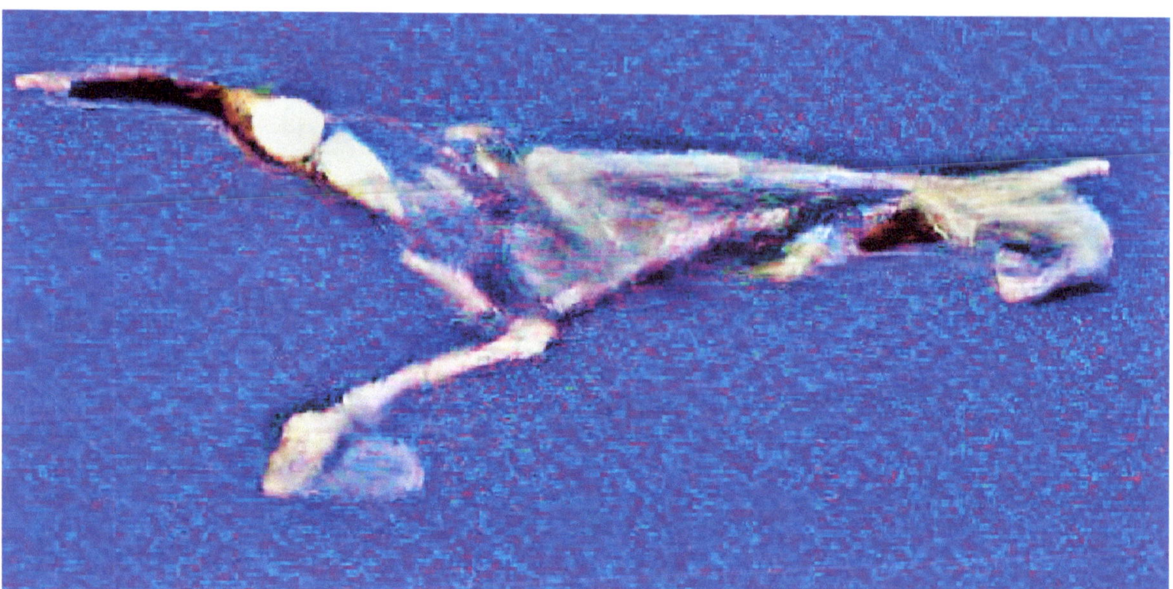

UFO Picture 1038 (Brightness Enhanced): April 30, 2019: 11:51:36 AM CST
Scott Gearen

When looking at this picture several thoughts come to mind, the first is the word "Plasma", is this what this is? It looks fluid, as if the entire object is blending together. Another thought is the shape, it appears the wing, if it is a wing, is presenting at a much higher angle than in previous pictures. The right side, which appears as the leading edge, slightly rounded with a black eye on each side, now has a round area not seen before on top of the object. The "capsule" now is shaped like a triangle, next to a "squinty" eye". Another thought... what is it?

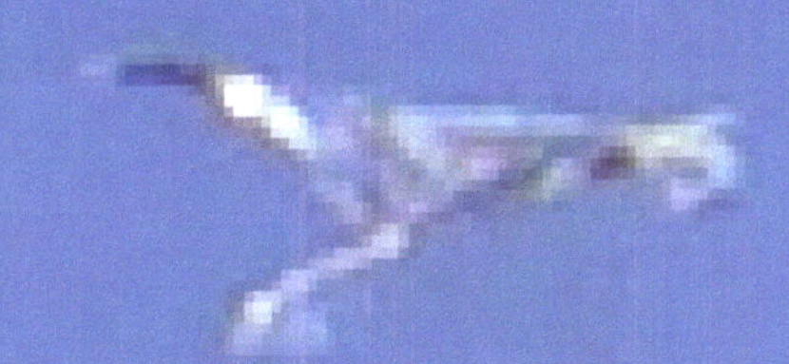

UFO Picture 1039 (Zeke Filter): April 30, 2019: 11:52:18 AM CST
Scott Gearen

Another part of the cycle where the object seems to be consumed by the bright areas over its entirety, it transforms to a bright sphere, disappears, and then reappears into some type of biological craft, that may have occupants. In the picture I could only see it with the aid of a filter.

UFO Picture 1039: April 30, 2019: 11:52:18 AM CST
Scott Gearen

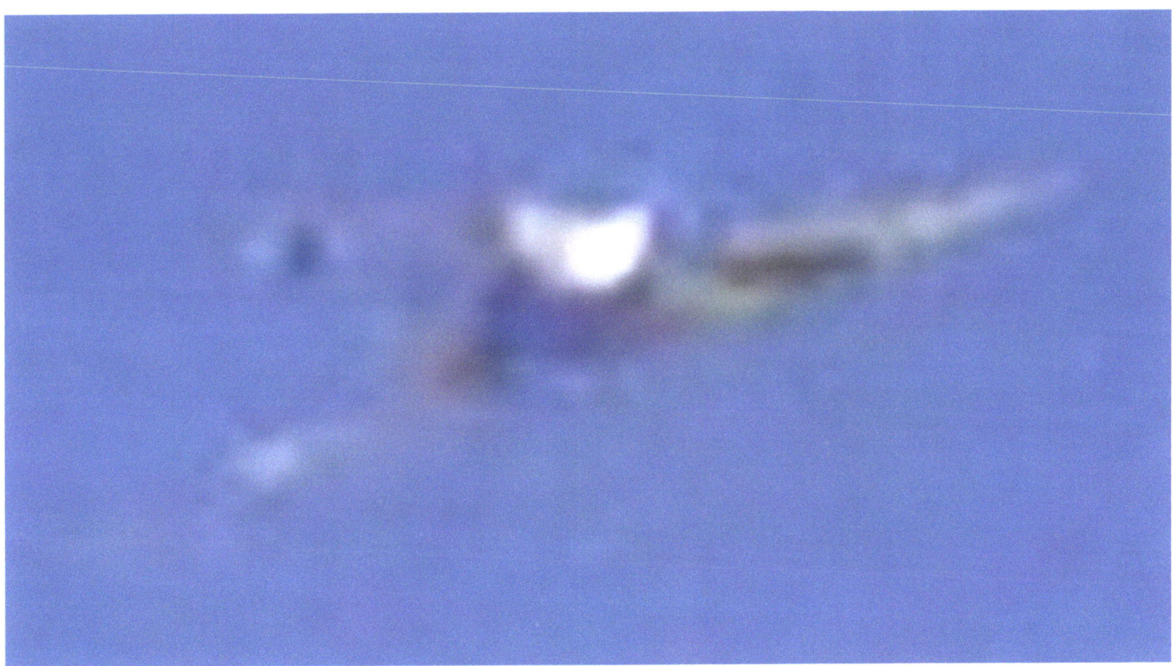

UFO Picture 1047: April 30, 2019: 11:56:18 AM CST
Scott Gearen

UFO Picture 1047: April 30, 2019: 11:56:18 AM CST
Scott Gearen

This time the new shaped object is dramatically different than previous transformations. The rounded head is now a distinctive point, the middle looks more like a compartment, and the rear area looks like two distinct tails trailing behind, similar to the swept back wings of a modern fighter jet. Filters will reveal a stunning revelation hidden within the object.

UFO Picture 1047 (Multiple Filters): April 30, 2019: 11:56:18 AM CST
Scott Gearen

My first thought was: this is a portal, or a window with muntins, but the changes exhibited by the object, made me think it could be a siphon for propulsion or gas exchange, akin to an octopus.

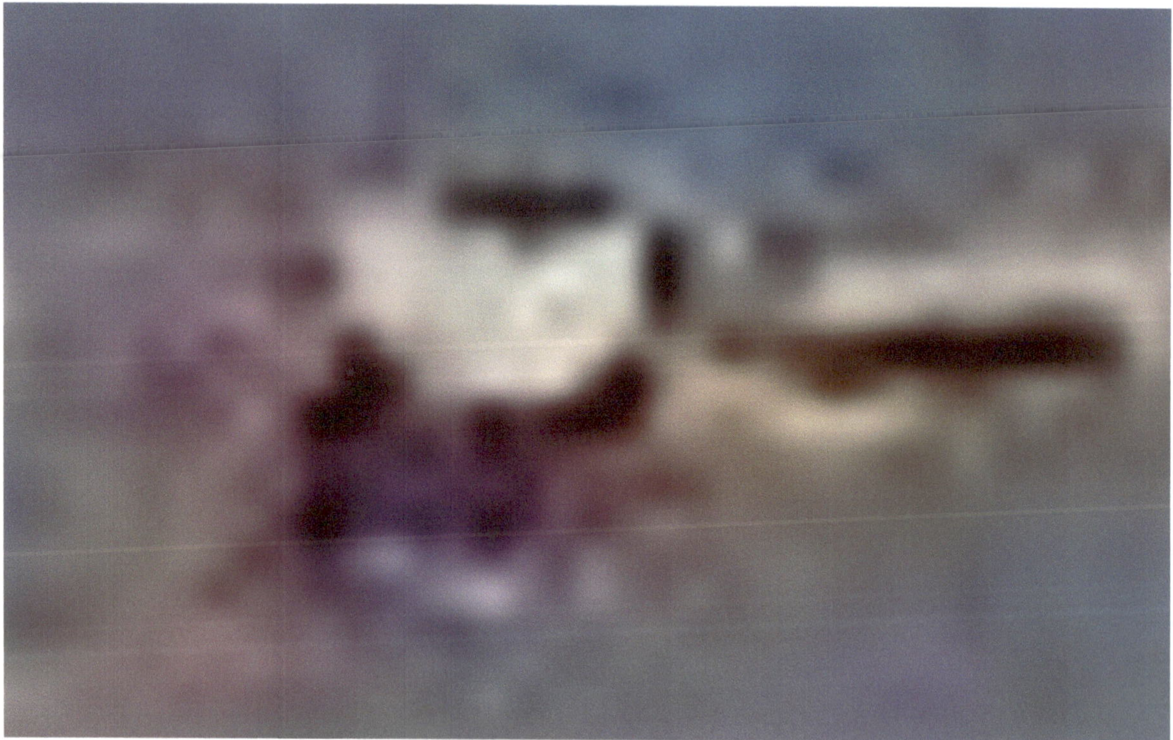

UFO Picture 1047 (Multiple Filters): April 30, 2019: 11:56:18 AM CST
Scott Gearen

UFO Picture 1048 (Zeke Filter): April 30, 2019: 11:56:24 AM CST
Scott Gearen

The previous two full page-pictures, taken less than a second apart, show the object is in the same location, and has quickly transformed from the distinctive shape to a sphere. The sphere in this picture was not visible to eyesight alone and was only detected with the aid of a filter.

UFO Picture 1048: April 30, 2019: 11:56:24 AM CST
Scott Gearen

UFO Picture 1051: April 30, 2019: 12:31:08 PM CST
Scott Gearen

At first glance this looks like a rocket traveling at hypersonic speed, maybe something closer to warp drive, but all the appearance of movement is within the object, it is not streaking across the sky, rather it has remained stationary in the same general area during the entire observation. Everything within the object appears to blend together during this transformation and then in just four seconds it will materialize into a completely new shape, (picture 1052) one that appears to look like an oceanic Manta-Ray, however, being that it is in the air, it is now a "Sky-Ray".

UFO Picture 1051: April 30, 2019: 12:31:08 PM CST
Scott Gearen

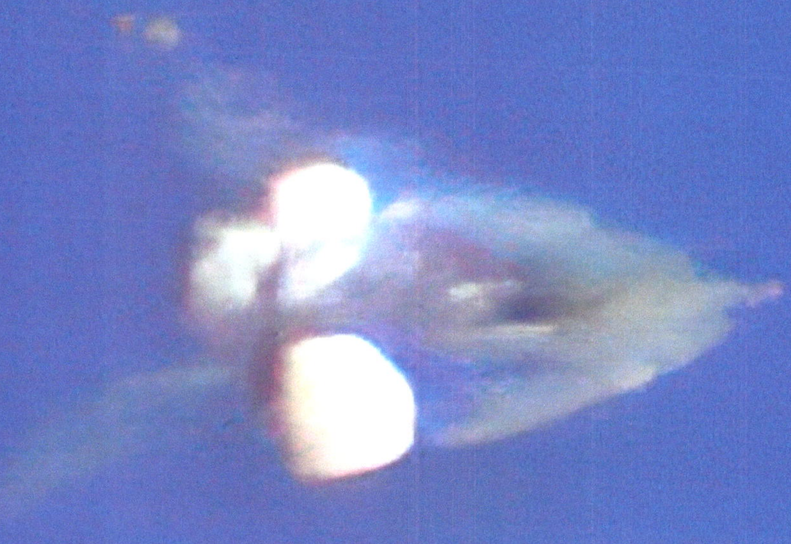

UFO Picture 1052: April 30, 2019: 12:31:12 PM CST
Scott Gearen

It is not common to see something like this above you, but it is accepted as normal if seen in the ocean, and identified as a Manta Ray. Why couldn't there be a species that is able to fly... I am convinced there is a species that is able to fly, swim, and travel throughout the atmosphere and beyond... "Transmedium". A large blue eye on the left side, common in oceanic species, is it real?

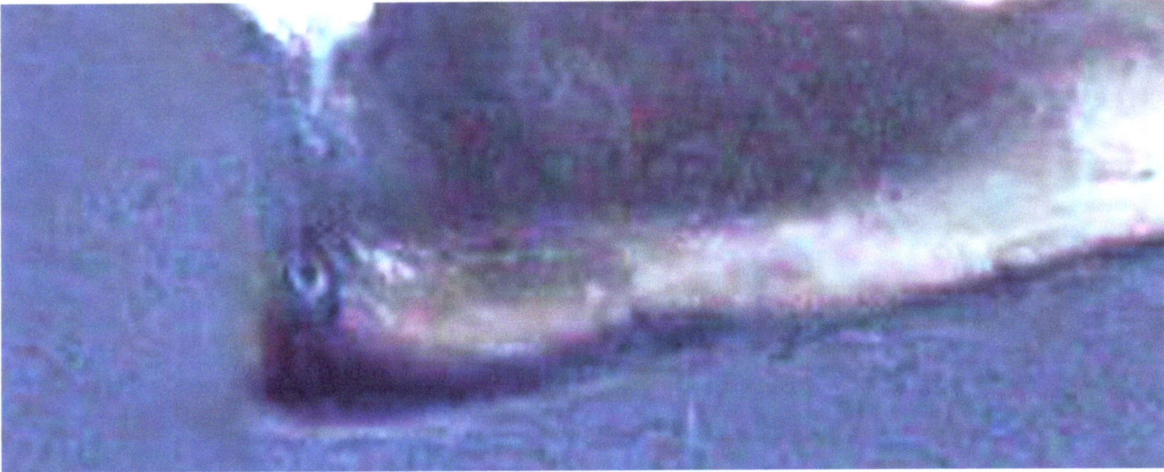

UFO Picture 1052: April 30, 2019: 12:31:12 PM CST
Scott Gearen

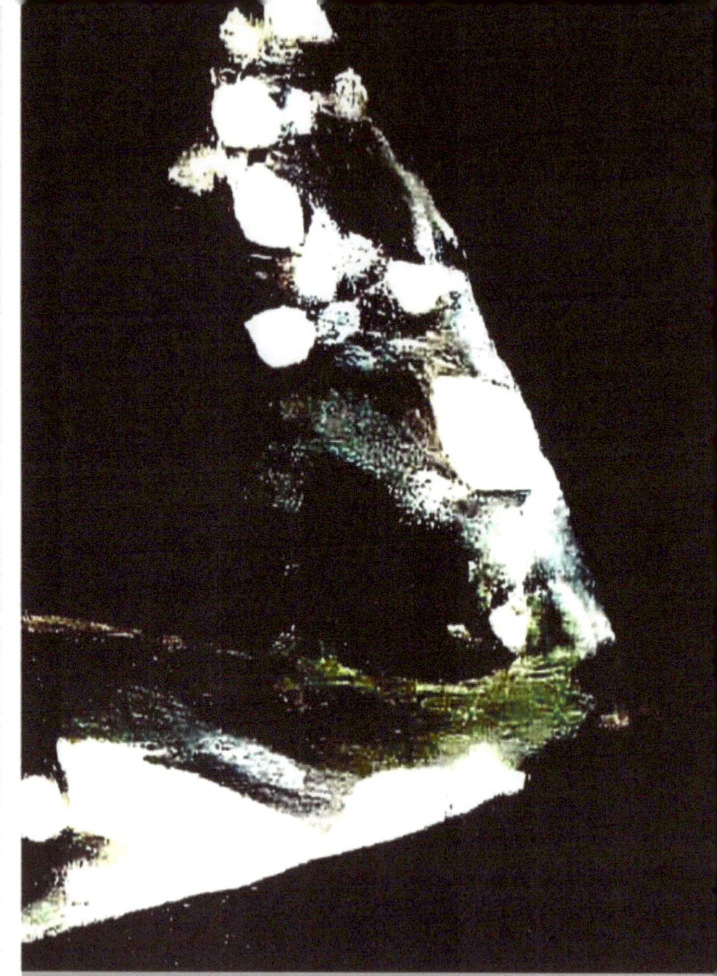
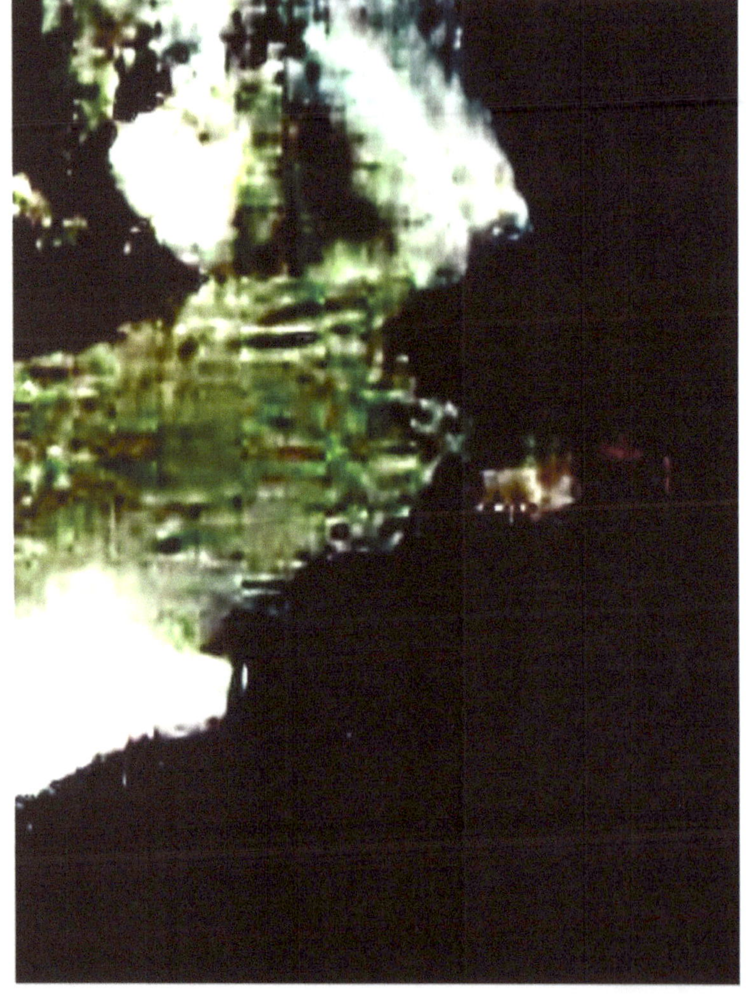
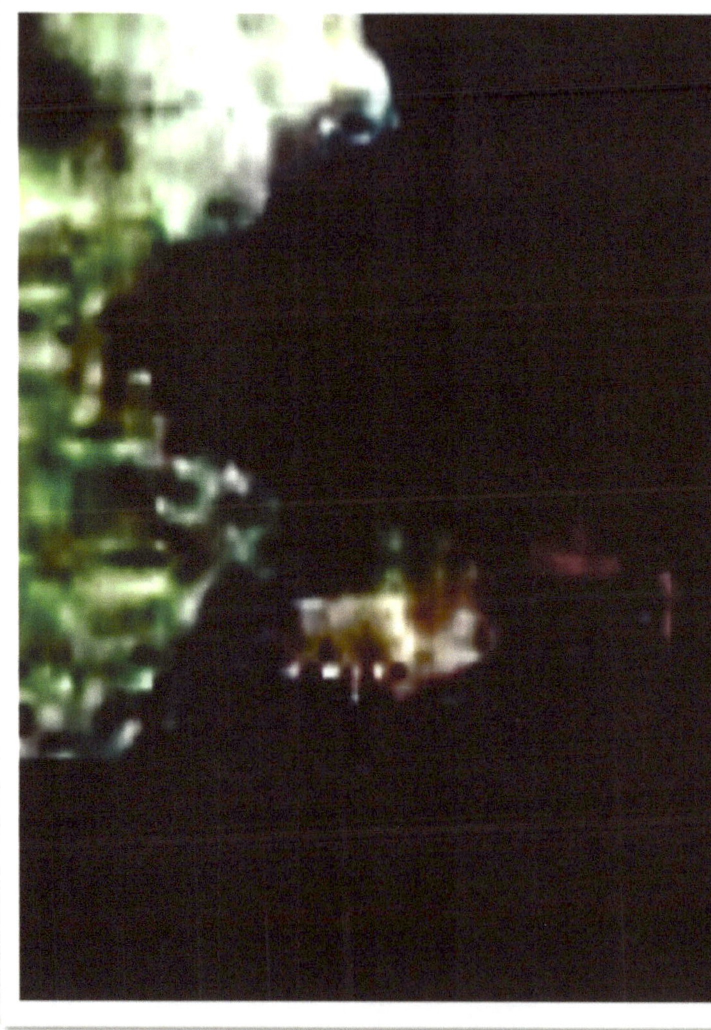

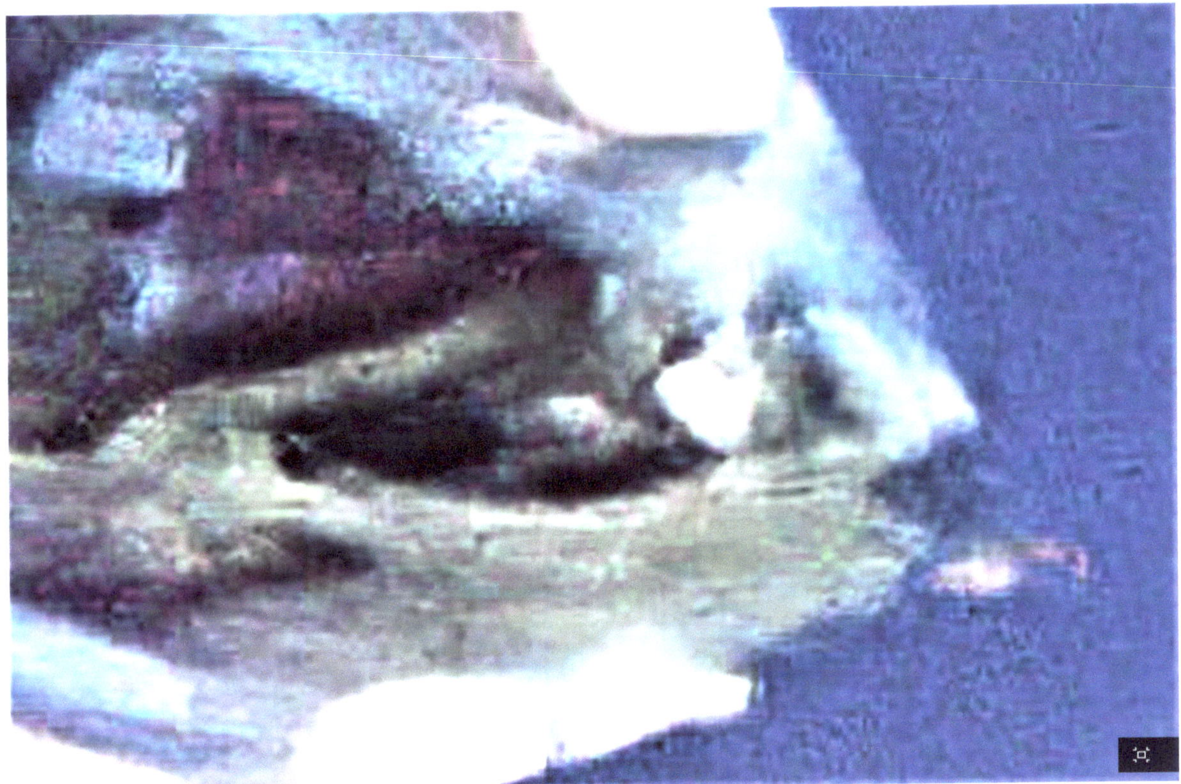

UFO Picture 1052: April 30, 2019: 12:31:12 PM CST
Scott Gearen

The opposite side of this object also appears to have a black eye, and an open mouth. It appears this object is either feeding or it is a mouthbrooder. Could this area be a nursery for this species?

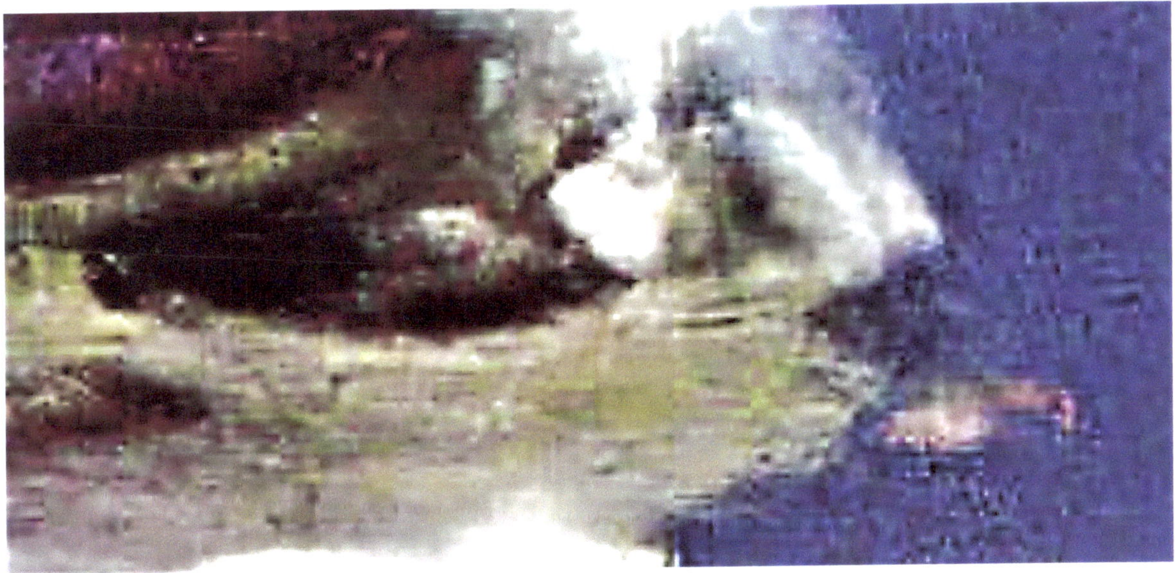

UFO Picture 1052 (Brightness Enhanced): April 30, 2019: 12:31:12 PM CST
Scott Gearen

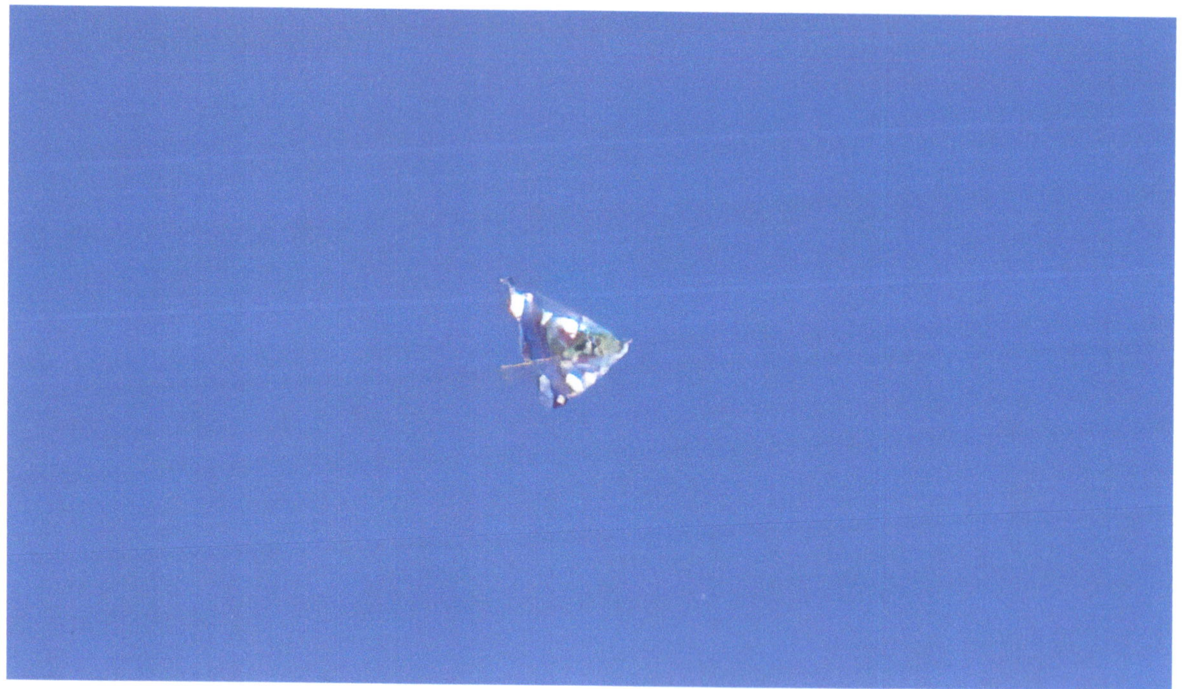

UFO Picture 1053: April 30, 2019: 12:31:18 PM CST
Scott Gearen

UFO Picture 1053: April 30, 2019: 12:31:18 PM CST
Scott Gearen

The object is still changing its shape, what is noticeable are the shapes and colors on the object itself. The large dark blue rectangular shape has been noticed often, as well as the black circular spot in the center, and a dark pyramid shape. The bright areas are always present and changing.

UFO Picture 1054: April 30, 2019: 12:31:26 PM CST
Scott Gearen

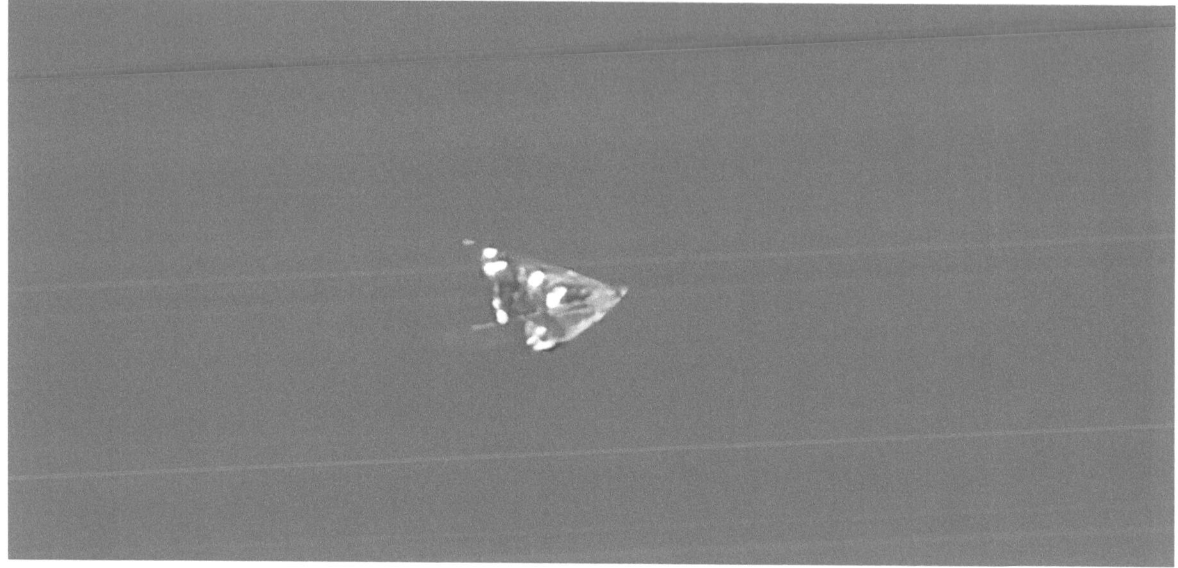

UFO 1054: April 30, 2019: 12:31:26 PM CST
Scott Gearen

A "Shape-Shifter", that is what a Native American told me after he took a look at these pictures. I believe him, this object continuously changed shape during the entire time I observed it.

UFO Picture 1055 (Zeke Filter): April 30, 2019: 12:33:48 PM CST
Scott Gearen

A filter identified as "Zeke" was applied to this picture to aid in locating the object, which now is a very small sphere. I was unable to locate it in the bottom picture without the aid of the filter and thought the camera lens without polarization had not been able to record it.

UFO Picture 1055: April 30, 2019: 12:33:48 PM CST
Scott Gearen

UFO Picture 1056 (Zeke Filter): April 30, 2019: 12:34:44 PM CST
Scott Gearen

Approximately 1:04 minutes later the object is still presenting as a sphere. Still unable to detect it without the aid of the Zeke filter to contrast the sphere against the blue sky. The small dark circles on the lens can be used as a landmark to locate the sphere and confirm its movements.

UFO Picture 1056: April 30, 2019: 12:34:44 PM CST
Scott Gearen

UFO Picture 1057: April 30, 2019: 12:34:50 PM CST
Scott Gearen

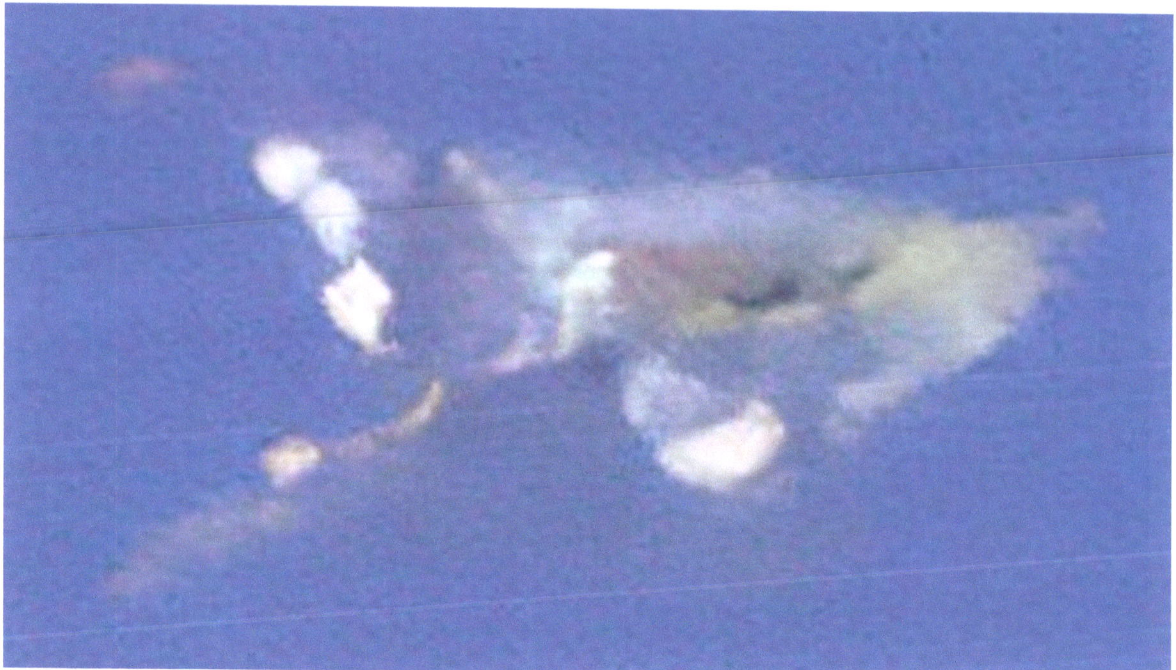

UFO Picture 1057: April 30, 2019: 12:34:50 PM CST
Scott Gearen

From a sphere to this object in just six seconds. Clearly the object is undergoing a transformation. A metamorphosis of sorts from one species to another... maybe; such as a caterpillar to a butterfly. Similar colors and shapes are present, the large dark rectangle, as well as the center brownish area. None of those transformations are created by humans, I don't believe this object has been created by modern humans. Is it something natural to the universe?

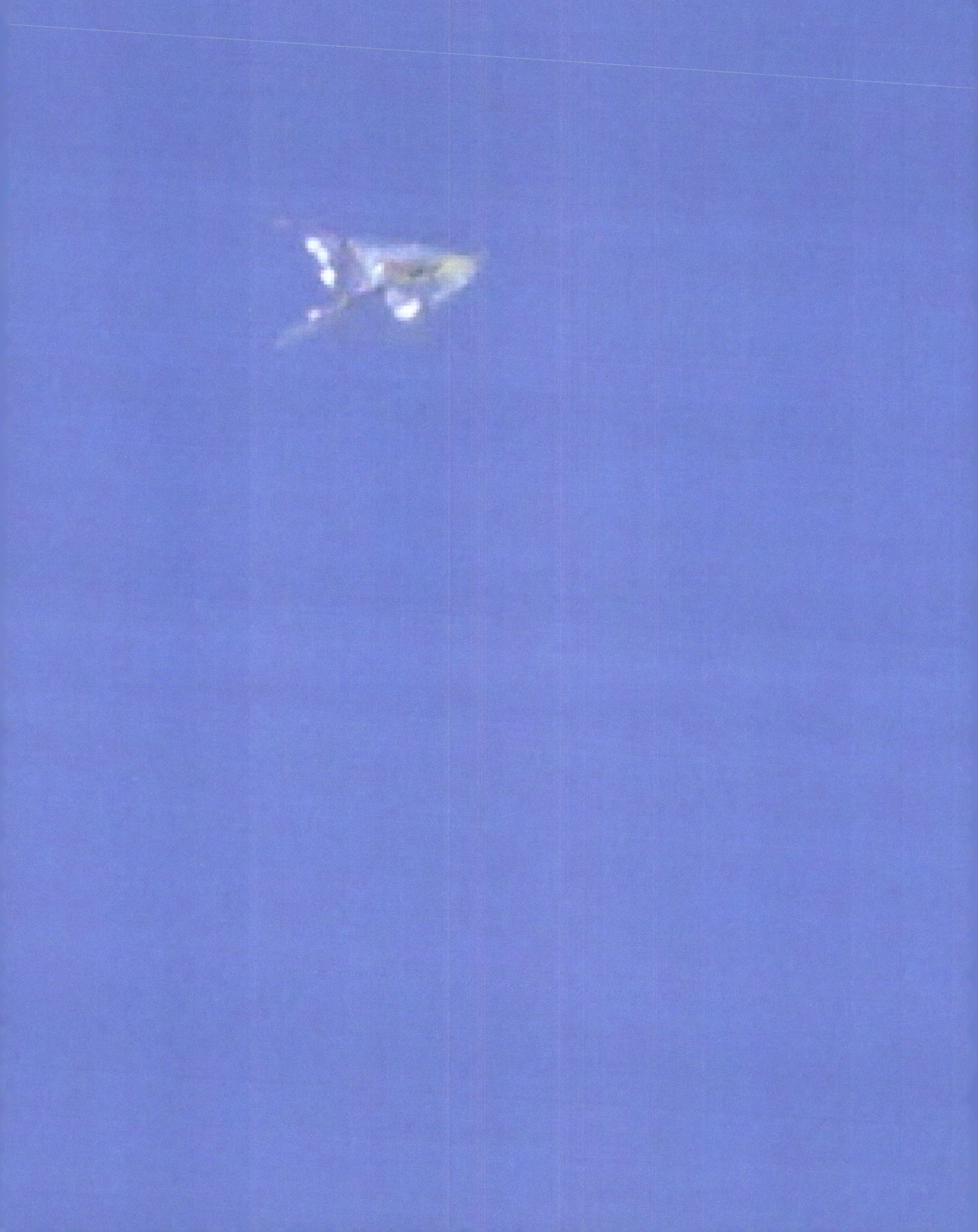

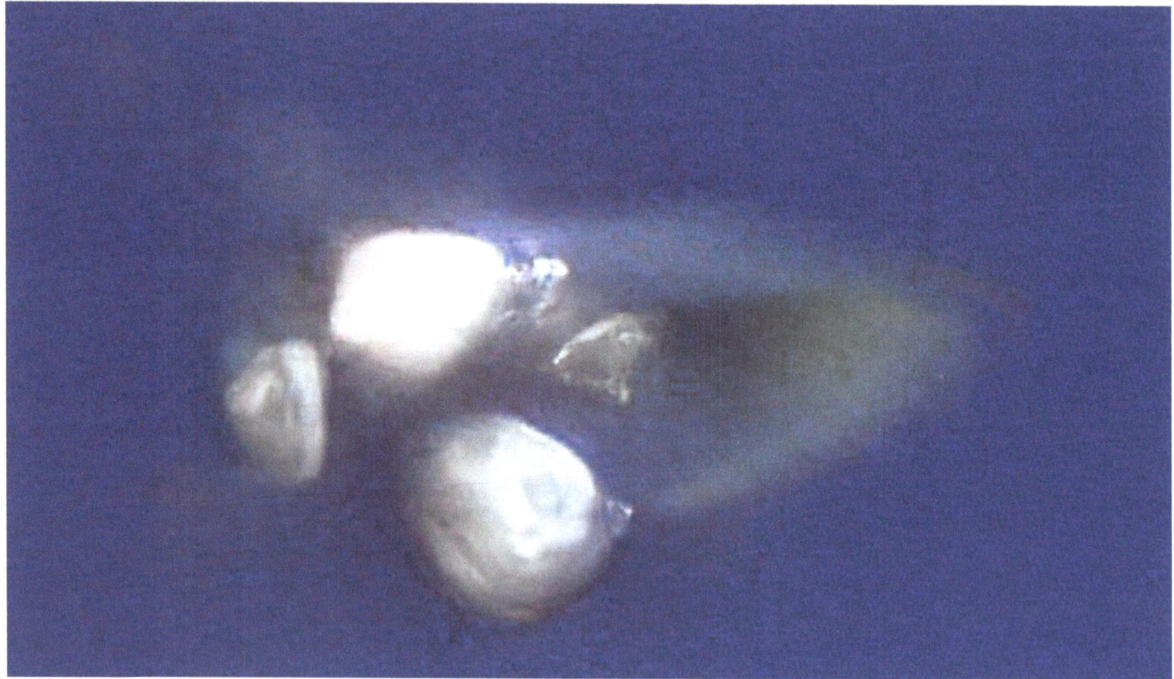

UFO Picture 1058: April 30, 2019: 12:35:00 PM CST
Scott Gearen

Ten seconds later 1058 resembles picture 1051, where it looks like a rocket traveling at hyper speed, but it isn't. It appears blurry without the speed to cause this effect. Is this a chrysalis forming inside a protective outer shell, like a cocoon for a caterpillar to become a butterfly... Is this transformation "holometabolism"?

UFO Picture 1058: April 30, 2019: 12:35:00 PM CST
Scott Gearen

UFO Picture 1059: April 30, 2019: 12:35:06 PM CST
Scott Gearen

UFO Picture 1059 (Enhanced Colors): April 30, 2019: 12:35:06 PM CST
Scott Gearen

Dramatic changes in six seconds. The object is solidifying and beginning to present in a new shape. Again, the beginning of a rounded head with eyes on both sides, and the rectangular dark area appears to be manufactured rather than something that would occur naturally, is it a capsule...

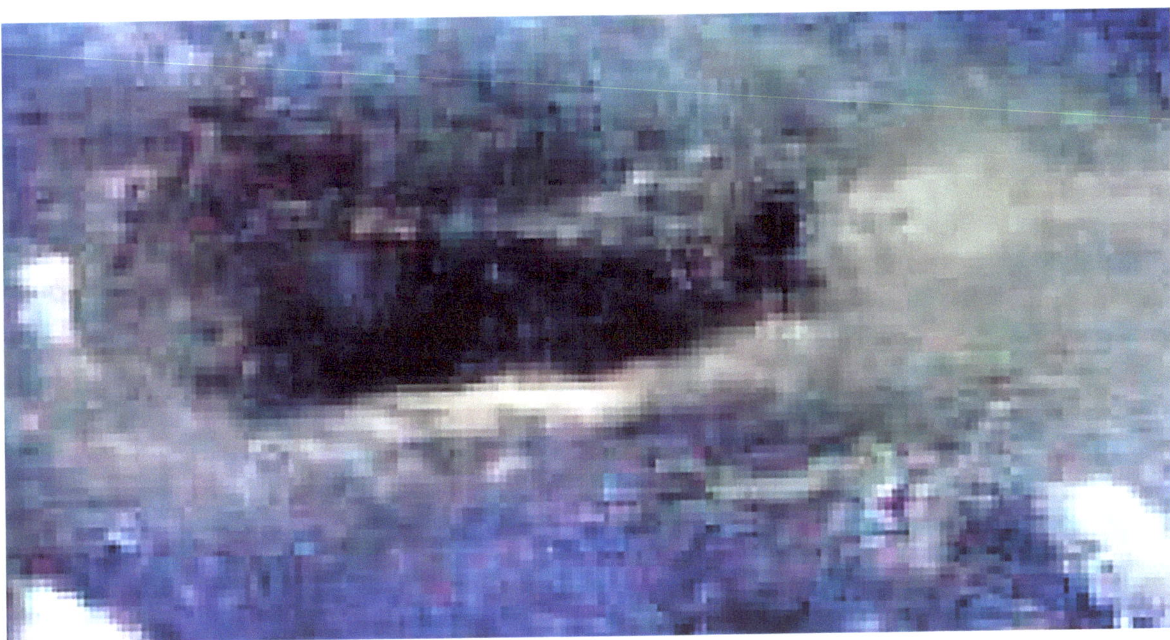

UFO Picture 1059 (Enhanced Color): April 30, 2019: 12:35:06 PM CST
Scott Gearen

The two pictures are enlarged with focus on the dark area I have referred to as a "capsule". The upper picture is color enhanced and the lower picture has a Sunscreen filter applied. Both pictures appear to show the dark area has depth to it. This is similar to picture 1035 where the Sunscreen filter was applied, and this one also appears to show a faint outline of a humanoid figure peering from within. Has Non-Human Intelligence evolved to travel within a biological craft?

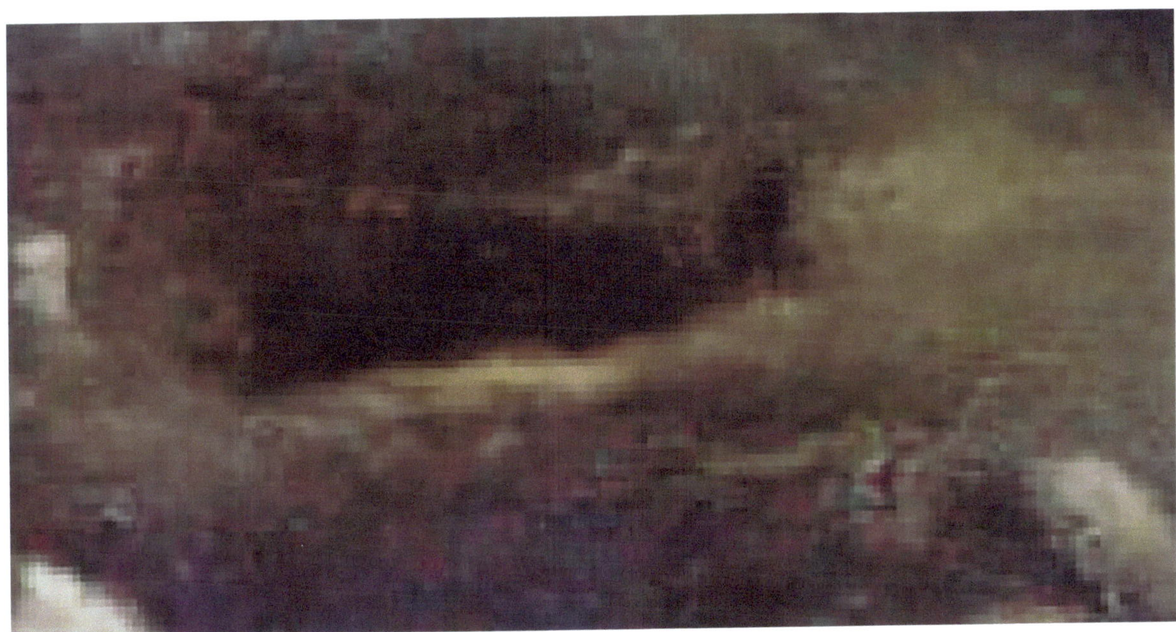

UFO Picture 1059 (Sunscreen Filter): April 30, 2019: 12:35:06 PM CST
Scott Gearen

UFO Picture 1060: April 30, 2019: 12:35:14 PM CST
Scott Gearen

There are many very interesting areas that stand out to me in this picture, the object seems to be higher on the far side and the top area is more exposed, many small bright circular objects can be seen, one of the bright areas is 'heart' shaped, which looks like the "heart" in picture 1052. The dark rectangular area is more defined, although it is difficult to determine if it has depth, and the straight edge across the top of the object does not appear natural.

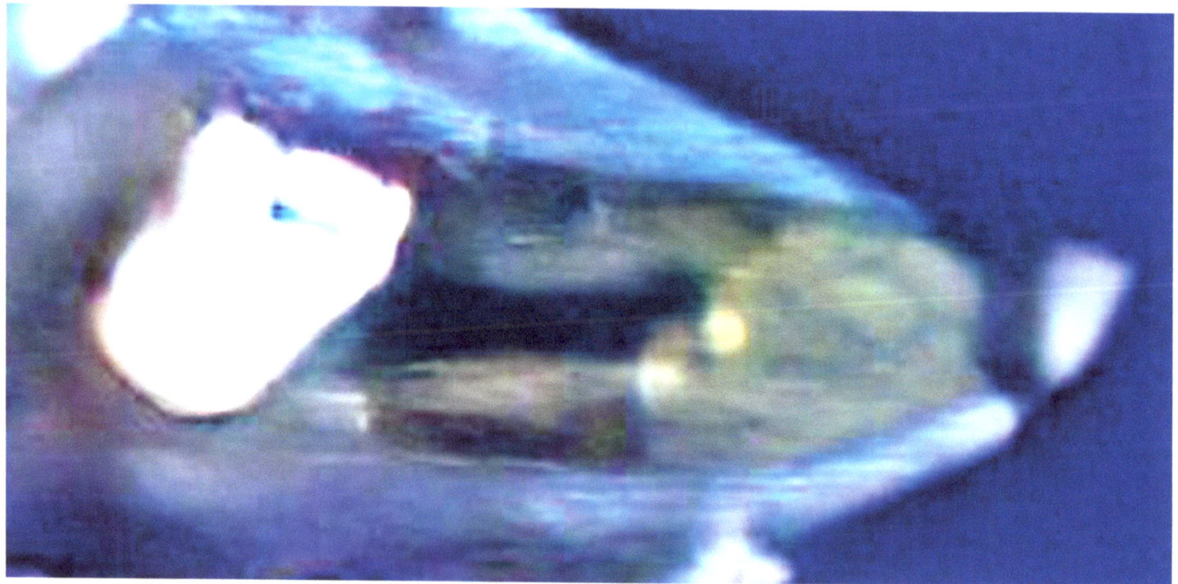

UFO Picture 1060: April 30, 2019: 12:35:14 PM CST
Scott Gearen

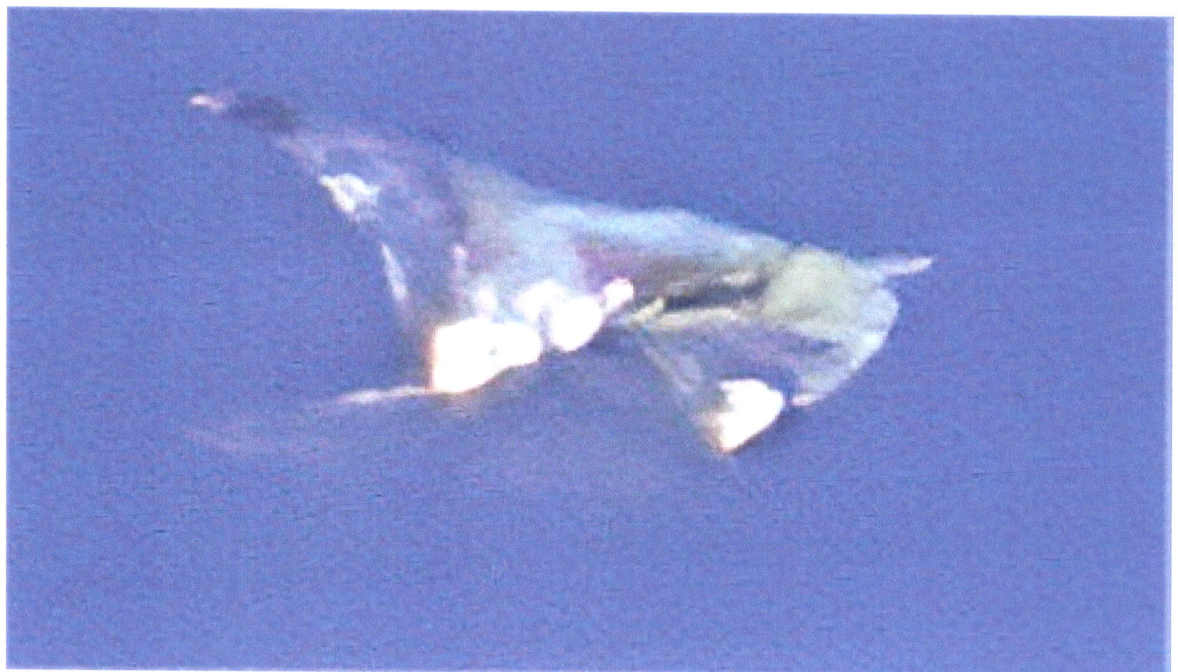

UFO Picture 1061: April 30, 2019: 12:35:30 PM CST
Scott Gearen

The right side appears to be a dramatic change and is now more rounded. I can't help but to think of the shape of a Bonnet Head Shark or even a Cownose Ray, both alien to humans, but accepted as a natural part of this planet. The small triangle on the right (picture 1060) now appears not only attached but maybe an eye-cover, just as the near black spot on the lower right appears to be. From the first time I saw this series of pictures I have always felt this object is more alive than a manufactured craft.

UFO Picture 1061: April 30, 2019: 12:35:30 PM CST
Scott Gearen

UFO Picture 1061 (Clarity Enhanced) April 30, 2019: 12:35:30 PM CST
Scott Gearen

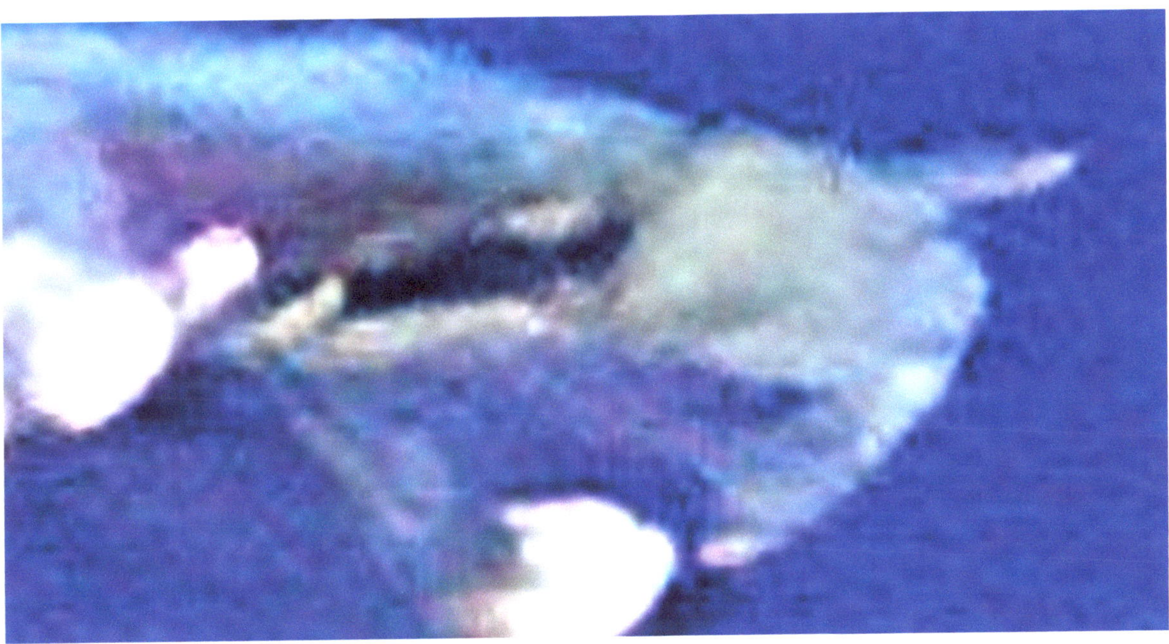

UFO Picture 1061 (Clarity Enhanced): April 30, 2019: 12:35:30 PM CST
Scott Gearen

Despite the changes that may appear to be a craft of sorts, my hypothesis is: I think this is not just the leading edge of a craft but it is also the head area of this species... a biological craft that an advanced non-human biological species has developed into a "Bioship". "Space Animal Hypothesis" may be more than just an educated guess.

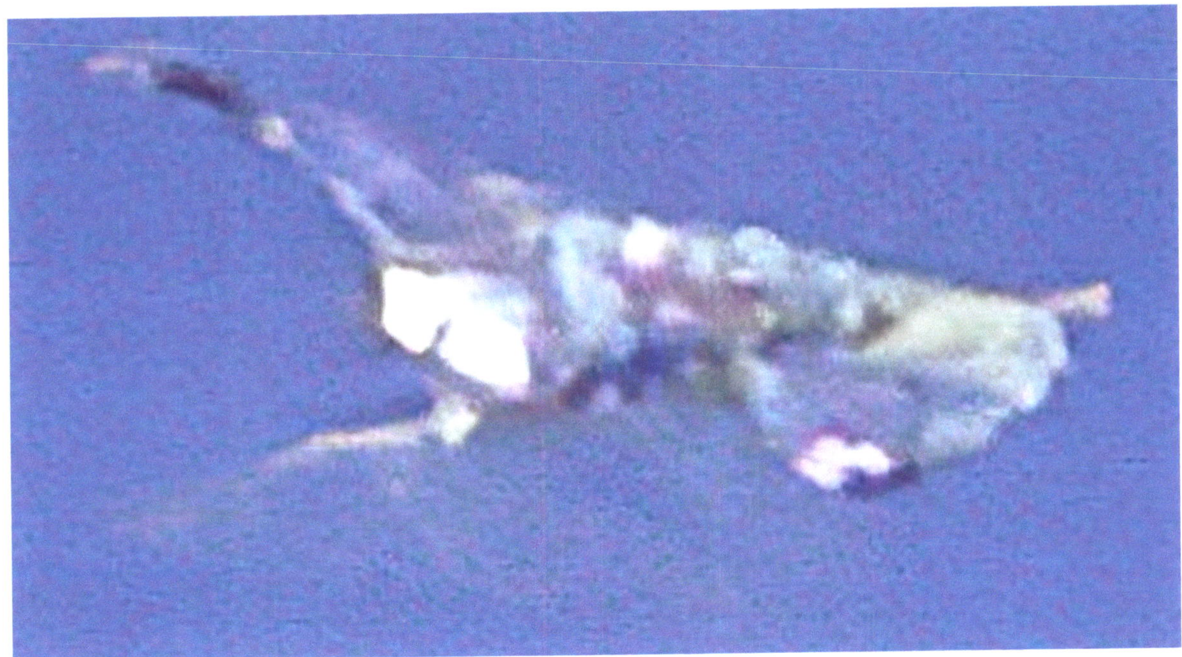

UFO Picture 1062: April 30, 2019: 12:43:36 PM CST
Scott Gearen

Another shape with bright areas that vary in size and intensity across the top. What appears to be an upturned wing with a blue underside and top side that can't be seen, although it may be a continuation of what can be seen. The center top area that has varied in a rectangular shape may be an opening to the interior. And a large black and white circular spot on the lower right area that resembles an eye, and something protruding on the opposite side could be another eye.

UFO Picture 1062: April 30, 2019: 12:43:36 PM CST
Scott Gearen

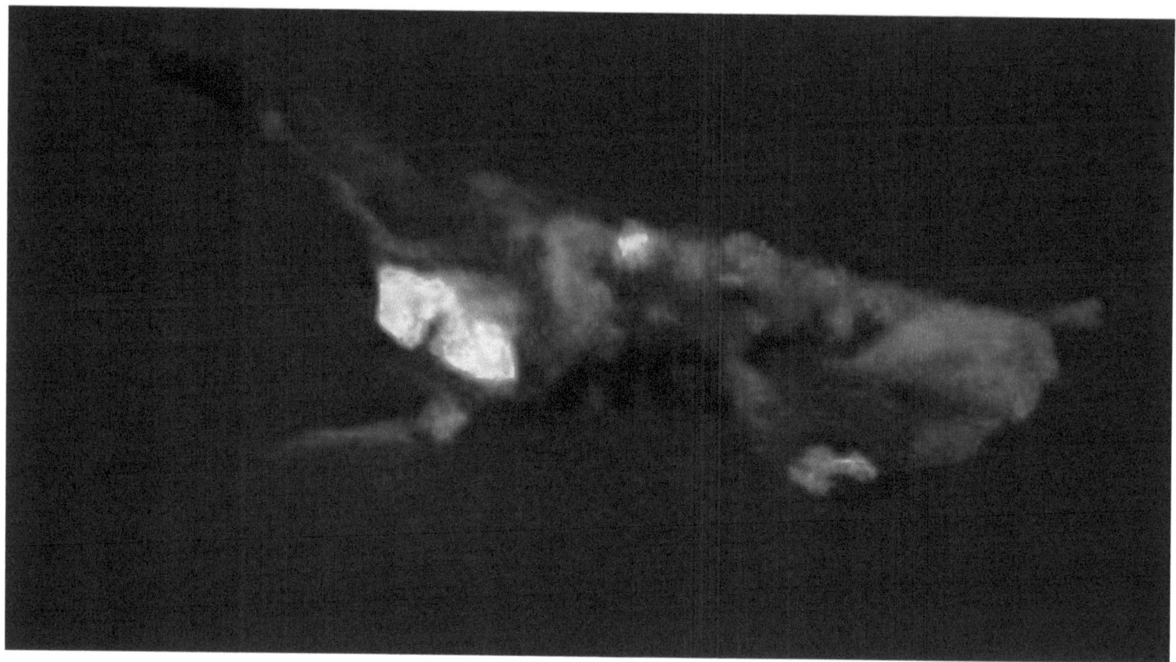

UFO Picture 1062 (Sunscreen Filter) April 30, 2019: 12:43:36 PM CST
Scott Gearen

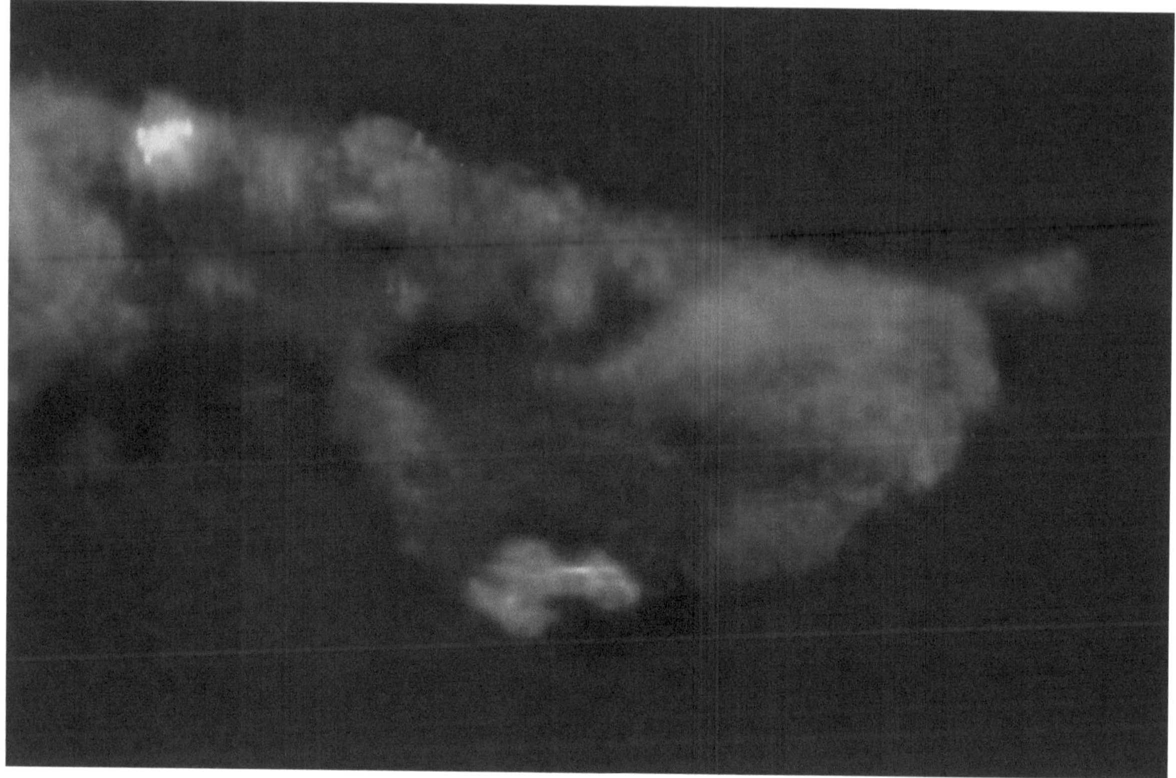

UFO Picture 1062 (Sunscreen Filter): April 30, 2019: 12:43:36 PM CST
Scott Gearen

Again, the image of a Bonnet head shark comes to mind, the edge is clearly rounded with distinct areas on each side where eyes are located on a Bonnet head shark, although without the wings, which is why it more closely resembles something from the cephalopod families.

UFO Picture 1063 (Zeke Filter): April 30, 2019: 12:44:08 PM CST
Scott Gearen

Whatever this object is doing, it is truly remarkable. Changing shape, making itself practically invisible, at least to human eyesight it is. The "sphere" seems to align with a cocoon, the same way a caterpillar creates a safe place for its metamorphosis into a new creature, except this object is able to change at a speed that is impossible for human technology.

UFO Picture 1063: April 30, 2019: 12:44:08 PM CST
Scott Gearen

UFO 1064: April 30, 2019: 12:44:16 PM CST
Scott Gearen

Very active phase with dramatic changes seem to be occurring. This is also the first time the far side "eye" is clearly identified, it is a very distinct blue circle and appears to have a yellow pupil.

UFO Picture 1064: April 30, 2019: 12:44:16 PM CST
Scott Gearen

UFO Picture 1064 (Sunscreen Filter): April 30, 2019: 12:44:16 PM CST
Scott Gearen

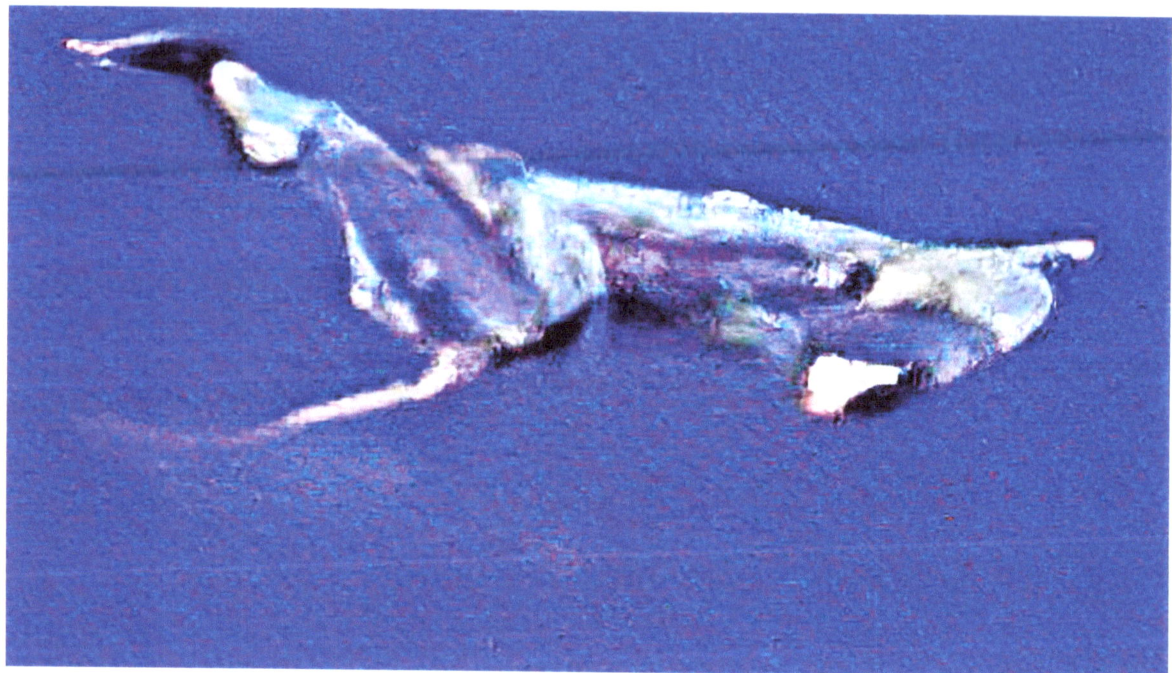

UFO Picture 1064 (Color Enhanced): April 30, 2019: 12:44:16 PM CST
Scott Gearen

Every picture seems more unbelievable than the previous one. Without knowing what this object is there is no way to know what it is doing, why it is doing what does, or where it comes from.

UFO Picture 1065 (Zeke Filter): April 30, 2019: 12:44:28 PM CST
Scott Gearen

During the last several pictures, the changes the object was going through was preventing me from seeing it with my eyesight, only with my polarized sunglasses was I able to see it. Conception to birth for all living things is very dramatic, could this be what is happening with this object? Nothing this object is doing is random, it is completely under intelligent control.

UFO Picture 1065: April 30, 2019: 12:44:28 PM CST
Scott Gearen

UFO Picture 1066: April 30, 2019: 12:44:36 PM CST
Scott Gearen

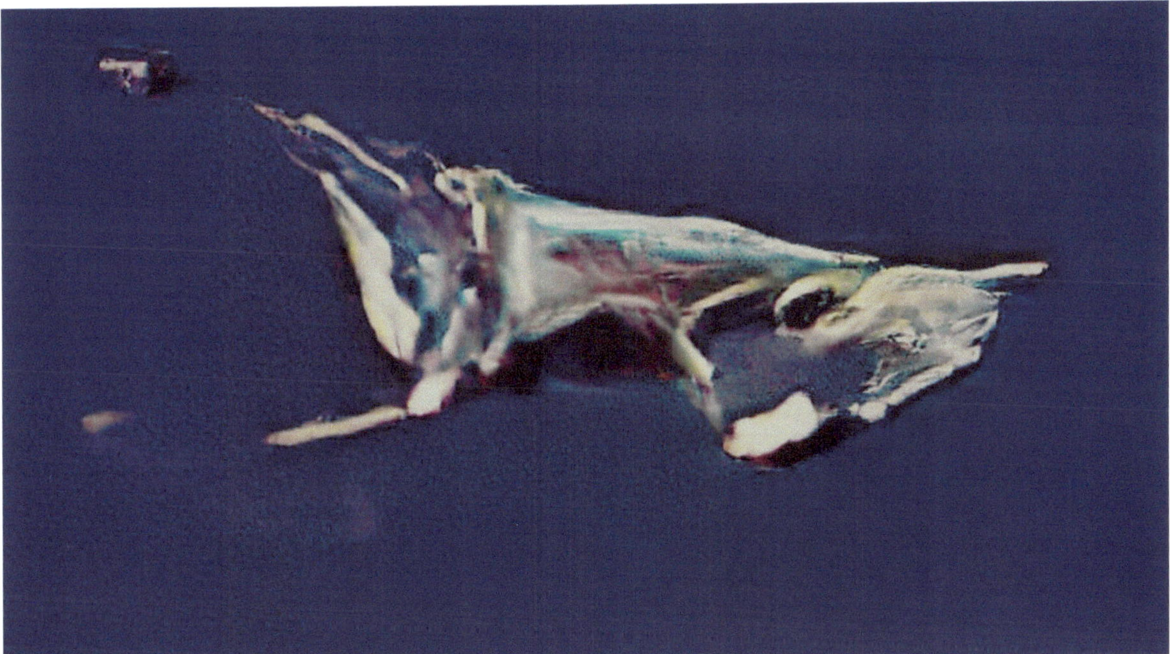

UFO Picture 1066 (Sunscreen Filter): April 30, 2019: 12:44:36 PM CST
Scott Gearen

What can change shape like this? Is this what has been reported as "plasma" in previous sightings? It resembles multiple liquids that are able to blend together, but how and why?

UFO Picture 1066 (Rogue Filter): April 30, 2019: 12:44:36 PM CST
Scott Gearen

Without an explanation as to what this object is, it becomes almost impossible to believe it is real. Much less to believe the camera was pointed at only one object that was able to change into so many different shapes within seconds, to disappear from human eyesight, change into a sphere, and then reappear in a completely new shape. This is what science fiction is made of… everything you knew as science fiction has become reality… and the reality is, this object, whatever it is, may be part of our world and we don't know it because it is hiding in plain sight.

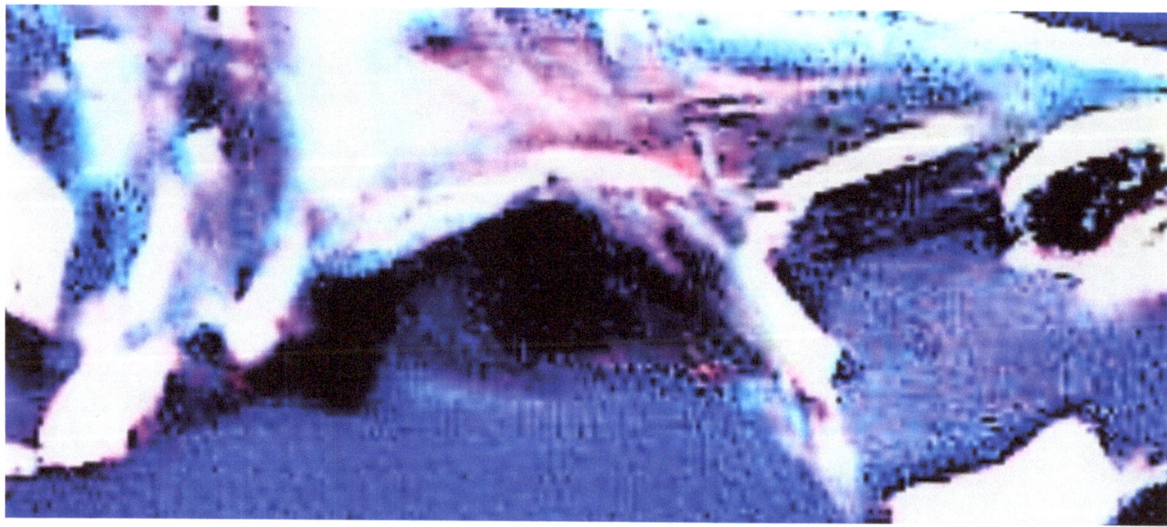

UFO Picture 1066 (Rogue Filter) April 30, 2019: 12:44:36 PM CST
Scott Gearen

UFO Picture 1067 (Zeke Filter): April 30, 2019: 12:44:46 PM CST
Scott Gearen

This sphere, that can't be seen in a clear blue sky (bottom picture) without the aid of filters, (top picture) could be in the atmosphere and never be noticed. A sighting like this has probably only been observed through electronics operated in an official capacity. A sphere, such as this would probably be reported as a craft rather than an Extraterrestrial Biological Being.

UFO Picture 1067: April 30, 2019: 12:44:46 PM CST
Scott Gearen

UFO Picture 1070: April 30, 2019: 1:12:36 PM CST
Scott Gearen

The pelican is what your eyes will focus on, you would never notice the object in the background. I was aware of the object, not the pelican and felt it was closer to me at this time than at anytime during the sighting.

UFO Picture 1070: April 30, 2019: 1:12:36 PM CST
Scott Gearen

UFO Picture 1070 (Zeke Filter): April 30, 2019: 1:12:36 PM CST
Scott Gearen

This is becoming a completely different shape than what has been seen in all of the previous pictures, except for one, picture 1047. Is this the culmination of the continuous transformations this object has been going through… "Kaiju" … maybe not just in movies…

UFO Picture 1070 (Golden Filter): April 30, 2019: 1:12:36 PM CST
Scott Gearen

UFO Picture 1083: April 30, 2019: 1:17:50 PM CST
Scott Gearen

After several hours, the object, which by now I felt a bond with, was leaving. Several more pictures were taken as it faded out of sight, but this is the last one I was able to see from my camera, however, I don't believe this is the last time we will meet. The pictures are not easy to explain.

UFO Picture 1083 (Zeke Filter): April 30, 2019: 1:17:50 PM CST
Scott Gearen

UF) Picture 1083: April 30, 2019: 1:17:50 PM CST
Scott Gearen

UFO Picture 1083: April 30, 2019: 1:17:50 PM CST
Scott Gearen

This is no ordinary Unidentified Flying Object, possibly never seen until now. A similar shape first appeared in picture 1047, without the amount of detail as seen here, but just as intriguing with the "siphon". Rather than a disc shaped saucer, this looks like a futuristic space craft, with an occupant observing me as I take pictures. The shaded area under the object, I think, is the movement of the craft, the object appeared so suddenly from a sphere the camera caught it as a blur.

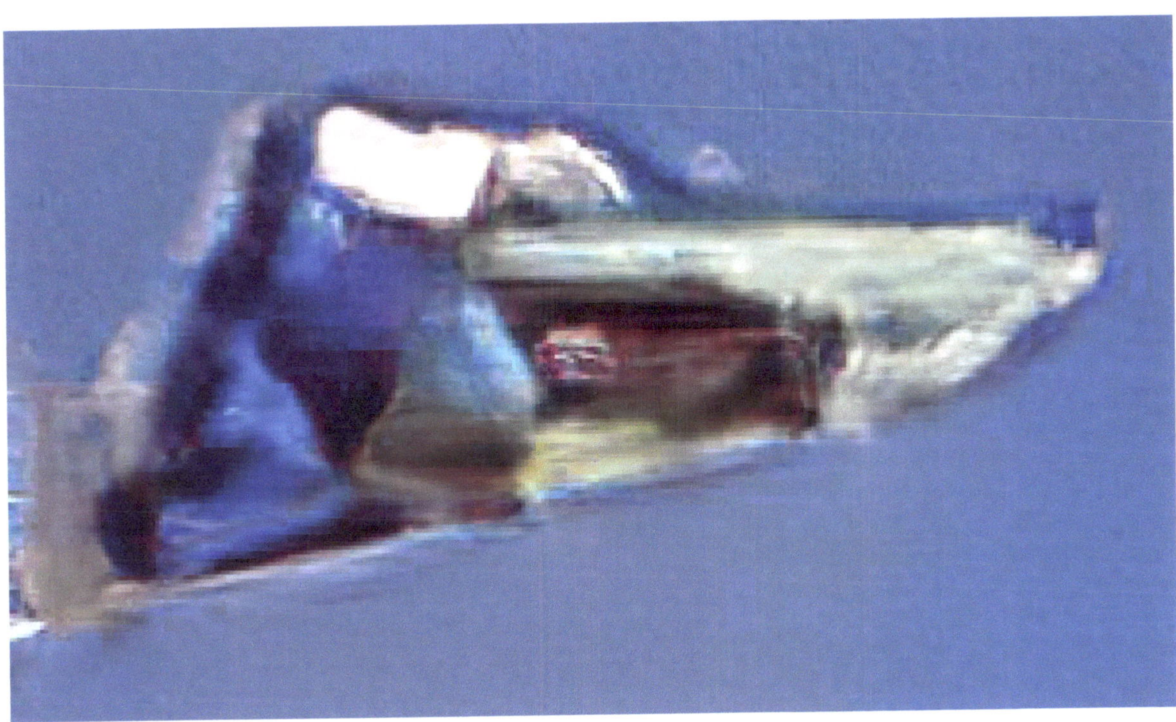

UFO Picture 1083: April 30, 2019: 1:17:50 PM CST
Scott Gearen

The focus here is on what appears to be a compartment within, and part of the object. This compartment is possibly the dark rectangle area that has been noticed in many of the pictures. There are many reports of humanoid figures, and there appears to be one in this compartment, possibly the pilot of this biological craft. Could this be from our future...

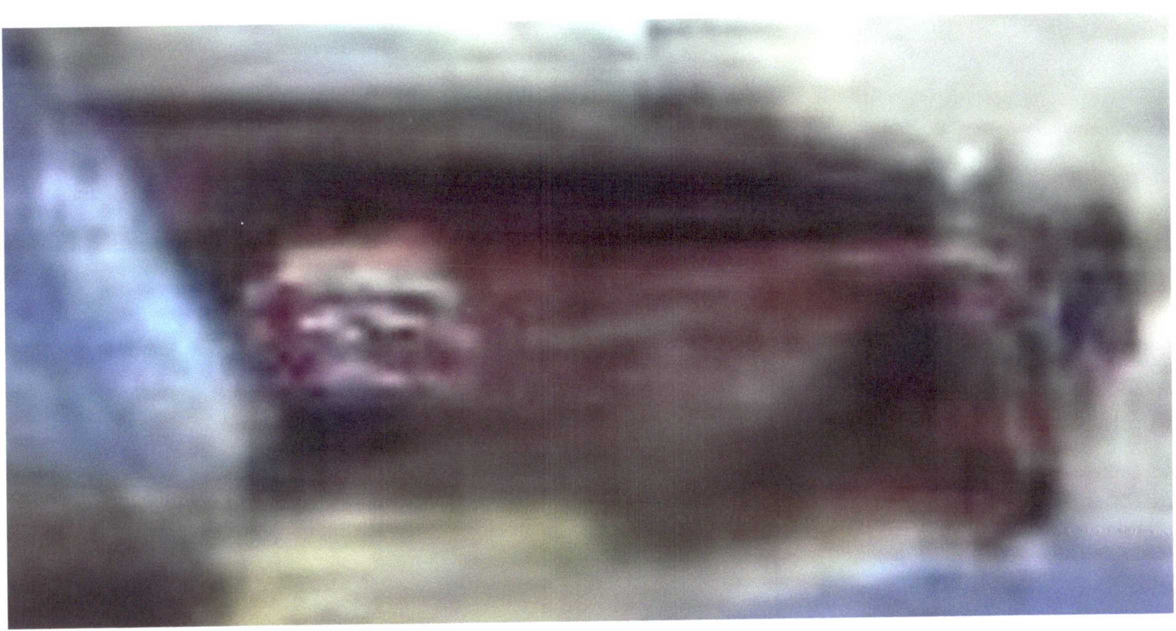

UFO Picture 1083: April 30, 2019: 1:17:50 PM CST
Scott Gearen

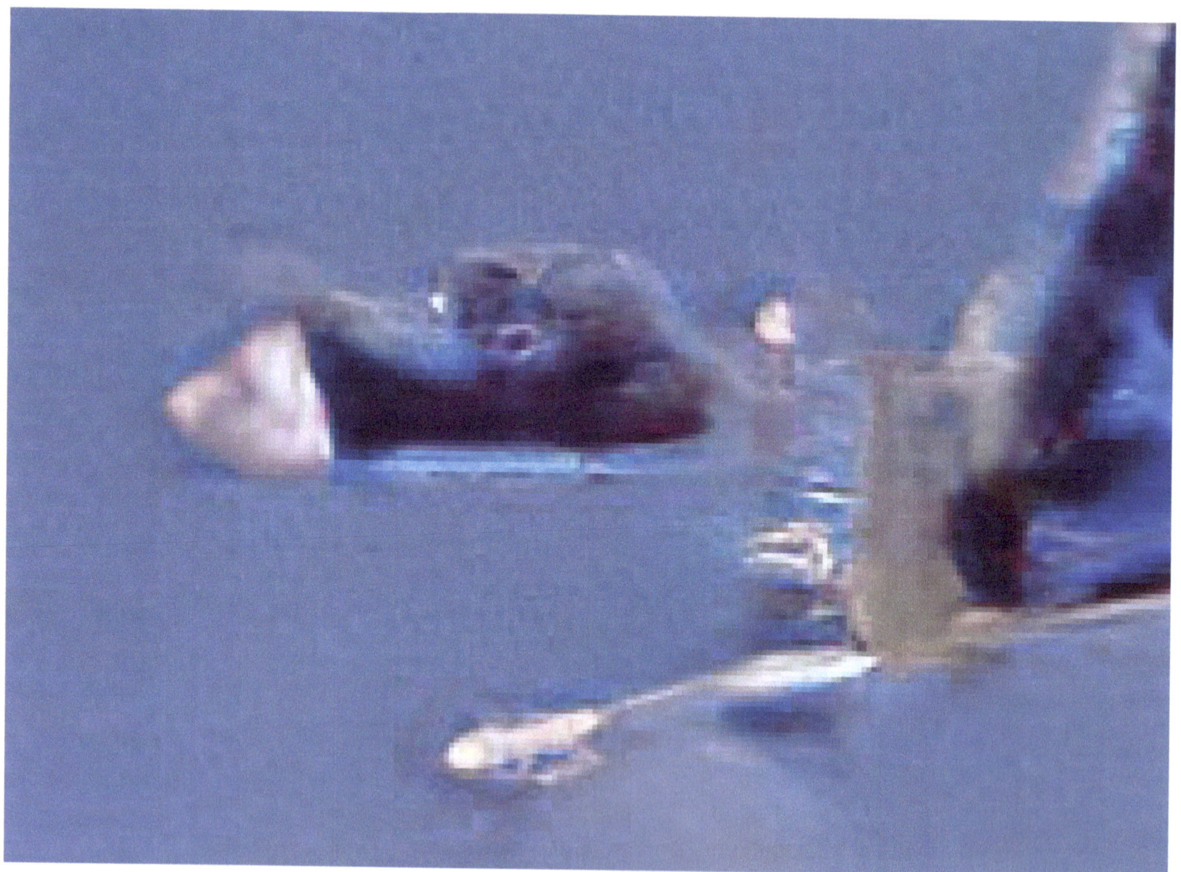

UFO Picture 1083 (Rear Capsule): April 30, 2019: 1:17:50 PM CST
Scott Gearen

Based on the aerodynamics of modern aircraft, I think this section is the rear area of this biological craft. Capable of being ejected and operating on its own, with occupants.

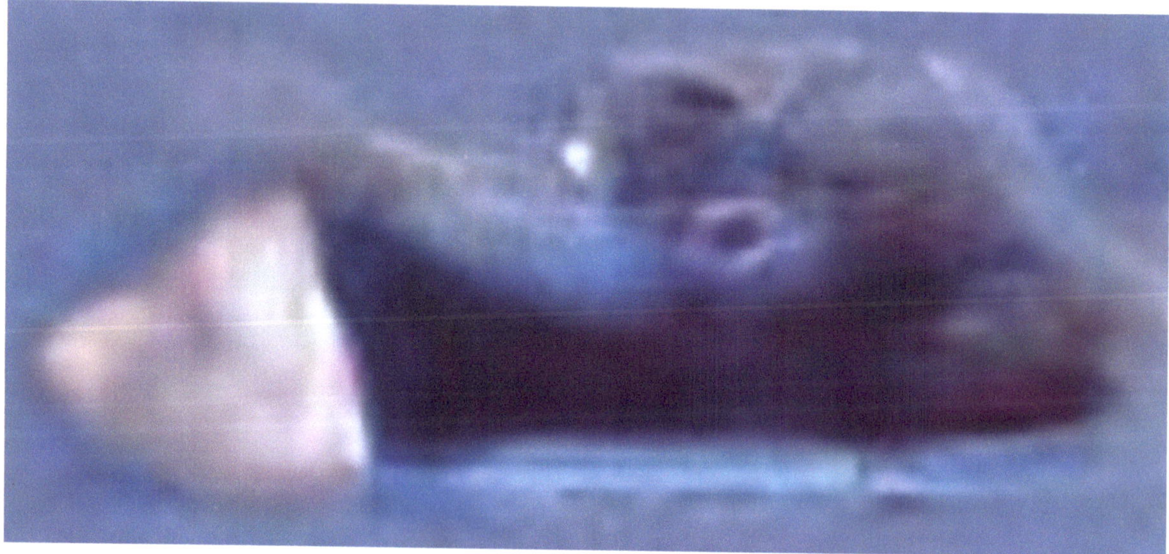

Picture 1083 (Rear Capsule): April 30, 2019: 1:17:50 PM CST
Scott Gearen

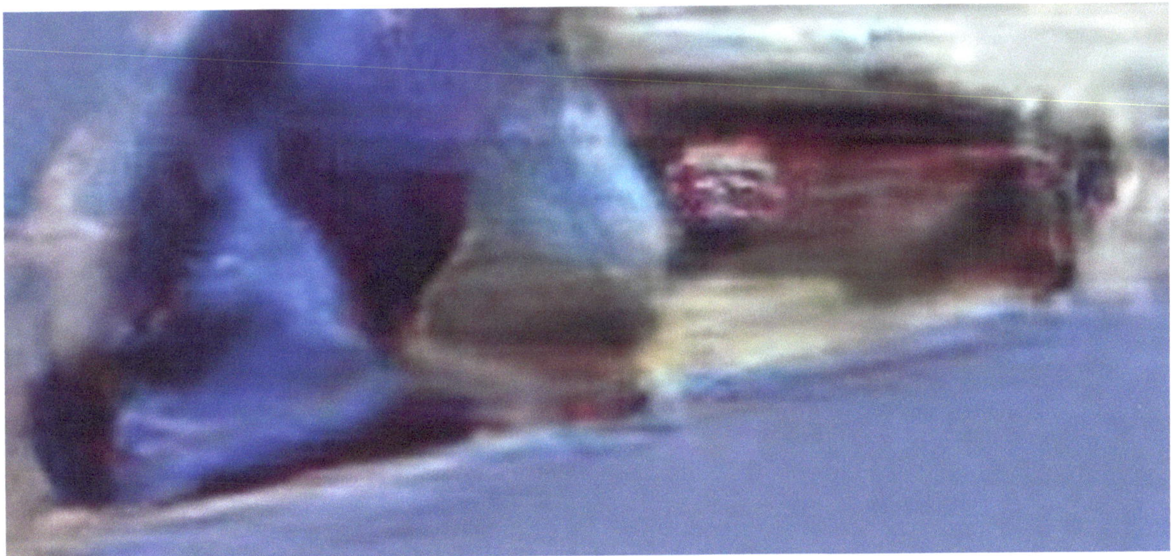

UFO Picture 1083 (Numbers): April 30, 2019: 1:17:50 PM CST
Scott Gearen

Hidden in the dark blue area could be clues leading to answers one day. The dark blue "pyramid" shape has been seen several times in previous pictures, each time in a different area. Close up of this triangle appears to have symbols across the top, most familiar is what is referred to as the number "8", followed by several symbols that could be real symbols or artifacts due to enlargement of the picture. A brown oblong shaped object has also been seen several times, with no guess as to what it may be. Maybe the occupant of this craft will reveal itself when they feel we are ready to understand who they are, and more importantly… who we are.

UFO Picture 1083 (Numbers): April 30, 2019: 1:17:50 PM CST
Scott Gearen

| three |

Are Reports of UFOs Taken Seriously

Why should we report sightings, and who should a sighting or an encounter be reported to? Should a sighting be reported directly to the Pentagon's organization established to collect data on Unidentified Flying Objects identified as "All-domain Anomaly Resolution Office"? Or another organization within the Department of Defense, such as the United States Air Force, United States Navy, or maybe the newly formed Space Command?

Many reported sightings are first reported to local Law Enforcement, but what do they do with these reports, do they provide this information to the Department of Defense or the Federal Bureau of Investigations (FBI)? What actions does an agency take when they receive a report of an Unidentified Flying Object? The government has a long history of organizations set up to investigate UFO reports, some that probably began before the famous Roswell incident in 1947.

After all the years since Roswell, and the many known and unknown government agencies investigating and studying Unidentified Flying Objects, there has not been official confirmation from the government of extraterrestrial crafts or beings, however, lack of confirmation doesn't mean they don't have proof, they just have not shared it with the public if they do. So, why report a sighting of an Unidentified Flying Object?

What about private agencies set up to receive reports from anyone that has had an encounter or sighting and wants to report it. How do you know who to report it to? Are some agencies better than others? Of the hundreds of thousands of sightings reported to multiple agencies in the United States and other countries, none have provided proof of extraterrestrial crafts or other Non-Human Intelligence. What would prevent a private organization being set up under a secret department of defense program to collect information on sightings that continue to occur on an almost daily basis? And if it really is a private organization, it is probably in the best interest of that organization to communicate and share knowledge with government agencies.

June 1947 was a very significant year for Unidentified Flying Objects; when Kenneth Arnold was flying in a small plane searching for a Marine aircraft that had crashed and been reported missing, he never expected to see "flying saucers". During his flight he reported seeing nine metallic saucer shaped objects flying near Mt. Rainier Washington at an estimated speed of 1200 miles per hour. He later described the objects as having the appearance of "saucers skipping on water", and the term was later referred to as "Flying Saucers" in newspapers. During investigations by the FBI, it revealed there were more witnesses to unidentified flying objects during that time than just Mr. Arnold.

In June 1947 a rancher in New Mexico discovered a field littered with an assortment of metallic debris on his property, which he later reported to local authorities, who in turn contacted the 509th Bomb Group Intelligence Office located at Roswell Army Air Field (RAAF). Maj. Jesse A. Marcel was assigned to investigate this incident. Soldiers from the base recovered much of the debris spread across the countryside, including a disc of unknown origin, and soon afterwards announced they were in possession of a "flying disc". This announcement created an international media frenzy when on July 8th the Roswell Daily Record made headlines stating the "RAAF Captures Flying Saucer on ranch in Roswell region".

Within a day of the announcement the Army retracted that statement that it was a flying saucer, and said it was a conventional weather balloon. At the time the RAAF was involved in a top-secret program identified as Project Mogul, which was conducted at multiple bases by the Army Air Forces launching high altitude balloons with listening devices to detect Soviet atomic bomb tests. Because of the need to maintain national defense secrets the change from a captured flying saucer to a weather balloon only created more suspicion about UFO's and ET's that has never gone away. There continues to be a question if it was a balloon or if it was extraterrestrial.

Jesse Marcel was the first military officer assigned to investigate the Roswell incident in 1947, and in 1978, after he had retired from the military as a Lieutenant Colonel he revealed to renowned ufologist, Stanton Friedman, that it was his belief the weather balloon claim had been a cover story, and his opinion was the debris was of extraterrestrial origin. Who is telling the truth... Lt. Colonel Marcel or the government?

Following Ken Arnolds sighting there were 100's of sightings from across the United States and other countries reporting unidentified flying saucer shaped objects, even a report from a United Airlines crew that they had witnessed nine unidentified flying objects similar to flying discs, which added to the credibility of the sightings. Mr. Arnold was first interviewed by officers from Hamilton Field in California. They concluded that Mr. Arnold had seen something he was unable to identify and would not have written a report if he did not believe what he had seen. The Army Air Force would publicly state the objects Mr. Arnold had witnessed was nothing more than a mirage.

Ironically at the same time the Army Air Force was conducting public interviews of witness's reported sightings of unidentified flying objects, members of intelligence units inside the Army Air Force, in conjunction with the FBI, began a secret investigation of the reported sightings. Based on this investigation it was concluded that the reports of "flying saucers" were not made up or easily explained and that there was indeed something real that was in the air and being seen. Based on this, General Nathan Twining would push for a formal investigation into these unidentified flying objects by other government agencies, which led to creation of Project Sign, also known as Project Saucer.

Project Sign was active for most of 1948 as an official U.S. government study of unidentified flying objects lead by the United States Air Force (USAF). In 1956 USAF Captain Edward Ruppelt claimed Project Sign had produced an "Estimate of the Situation" regarding Unidentified Flying Objects, with an explanation that UFOs were interplanetary in nature. General Vandenberg would later shutdown Project Sign stating lack of proof.

Although Project Sign was shut down, sightings of unidentified flying objects continued and Project Grudge was established to continue investigations into unidentified flying objects, but primarily to reduce the public concern over these unidentified flying objects. Officials decided to terminate Project Grudge to prevent panic in the populations, they felt having a program to investigate Unidentified Flying Objects would encourage people to believe there were actually objects in the atmosphere that were unidentified, and in December of 1949 it was shut down.

Continued sightings, as well as continued interest by the United States Air Force, led to the creation of Project Blue Book. Probably the most widely known official organization the general public was aware of that was set up to investigate reports and determine the validity of UFO's. Project Blue Book was established at Wright Patterson Air Force Base (WPAFB) March 1952, several years after the Roswell UFO incident, and would continue until December 1969. Captain Edward J. Ruppelt was named as the first director of Project Blue book. Rumor and speculation from the Roswell incident have it an alien spaceship or two crash landed in the New Mexico desert, and there were alien bodies recovered, some were buried and some sent to an unknown location, speculation is they were sent to the highly secret Area 51 in Nevada, and/or to the location where Project Blue book (named after "blue" booklets used at universities for testing) was established: Wright Patterson Air Force Base in Ohio.

An interesting fact was WPAFB was the Head Quarters for the Foreign Technology Division of the USAF. Something like a crashed craft from an unknown origin, possibly from another planet or solar system somewhere in the universe, another dimension, or even within the earth would certainly fall under "Foreign Technology", maybe "foreign technology" is not limited to countries outside of the United States. Of note: WPAFB, is home to the National Air and Space Intelligence Center, this organization began in 1961 and grew from the Foreign Technology Division of the 1940's.

Project Blue Book did not receive its name because of the blue uniforms worn by the Air Force, rather its name was in reference to "blue" booklets used in some colleges for testing. College exams were a serious matter and it was said by Ruppelt that high ranking officers were serious about this new project to investigate Unidentified Flying Objects.

With no formal agency for people to report their sightings many were reported to law enforcement or to the Air Force. Each U.S. Air Force Base had a Blue Book officer that would collect UFO reports and forward them to Project Blue Book. By the time Project Blue Book was shut down they had collected over 12,000 Unidentified Flying Object reports. Most of these sightings were identified but over 700 sightings were unexplainable. The scientific consultant on Project Blue Book was famed Astronomer J. Allen Hynek, who had also worked under the preceding projects: Sign and Grudge, and is credited with categorizing the types of sightings and encounters people were having, which continues to be used to categorize encounters. This categorization was made common knowledge in the movie "Close Encounters". Initially Mr. Hynek was skeptical of unexplained flying objects but was more open to the possibilities of what the unexplained flying objects actually were as he investigated reports and was unable to identify everything reported. Project Blue Book was terminated on 17 December 1969 following the Condon Report, which was the product of a committee funded by the United States Air Force at the University of Colorado from 1966 to 1969. This committee, under the direction of Edward Condon came to the conclusion that further study of the UFO reports was probably not going to produce any scientific discoveries.

The Air Force summarized the termination of Project Blue Book investigations into Unidentified Flying Objects by stating they had not found any of the UFOs reported to be a threat to national security, that there was no evidence that supported unidentified flying objects were anything more than what was already known by modern science, and there was no evidence of any of the unidentified flying objects being extraterrestrial. What the Air Force didn't say was investigations of UFOs would continue under different names and different organizations within the government. Many years later it would come to light those investigations into UFOs never stopped.

When one door closes, another door opens, just because the official USAF organization Project Blue book was terminated did not mean the sightings, and close encounters would shut down as well. Sightings continued to be reported, regardless of lack of knowledge that there was an organization investigating the sightings, the reports continued to occur. Coincidently, or is it a coincidence, the same year, 1969, that Project Blue Book was terminated, a private organization identified as Mutual UFO Network (MUFON) was established, was this a coincidence? MUFON is now one of the oldest and largest civilian organizations with over 4000 members, across the United States, and more than 40 countries around the world.

Sightings of unidentified flying objects continued and now because the Air Force stated they were no longer investigating UFOs, there was a void to fill and private organizations became a way for people to report sightings. The data collected by private organizations was a gold mine of information that would be too important for intelligence organizations to ignore. It is not too far-fetched of an idea to consider these organizations would be monitored by government agencies to stay abreast of possible threats to the nation. Rather private organizations were set up secretly by government agencies or citizens with an interest to identify what was being seen, this collection of data could be used by government agencies to continue to investigate unidentified flying objects without scrutiny from the general public as well as adversarial nations?

If there was no evidence of Unidentified Flying Objects, why would there be a need for the Army, Navy, and Air Force to create Joint Regulation number 146? Created in 1953, Joint Regulation 146 made it a crime for any military personnel to discuss classified UFO reports with unauthorized individuals. If someone was found guilty of violating this regulation they would face up to two years in prison and or be fined up to $10,000. Certainly, a good reason for anyone with information about UFOs to keep it secret.

Where would a classified UFO report come from? What would make a report classified? Since this is a Joint military regulation directed at classified UFO reports, it has to imply that these reports most likely come from military or government personnel. But, could a report submitted from outside a military organization, from the general public, become classified? I believe it is possible, it can't only be pilots that have encounters and sightings of Unidentified Flying Objects. Having a Joint Regulation in place, where someone would face jail and monetary fines has probably been the reason that members from military organizations, and others associated with Top Secret clearances do not talk about sightings, especially an encounter such as mine... Hiding In Plain Sight, and Hiding In Plain Sight Volume II.

In 1969 Project Blue book came to an end, the United States Air Force said they were no longer investigating unidentified flying objects, and it was assumed that was the end of the Department of Defense collection of information and investigations into UFOs. But, was that really the end or was it just the end of the Air Force being the known component agency to compile and investigate this matter... or did the Department of Defense take these sightings to be so important to National Security that investigating and determining what these unknown objects were the investigations would continue under multiple layers of secrecy and headed by unknown and unidentified agencies to prevent anyone from knowing just how real and how seriously this matter could be to National Security...

Fast forward to 2017: The story published in the New York Times exposed there were ongoing Department of Defense investigations into unidentified flying objects. At least as far back as 2007 the Department of Defense had programs in place to investigate unidentified flying objects. Surely there were programs in place from 1969, when Project Blue book ended, to 2007 to investigate and determine what the unknown objects being seen and reported actually were, National Security is too important to ignore unknown objects in the air space over the United States.

The story in the New York Times, written by Helene Cooper, Ralph Blumenthal and Leslie Kean, provided evidence that Senator Harry Reid, as the Senate Majority leader, had pushed for funding of a secret organization identified as the "Advanced Aerospace Threat Identification Program." (AATIP). Which implies to me that their mission was to determine if Unidentified Flying Objects posed a threat to National Security.

Department of Defense Pentagon statements indicated AATIP had run out of funding by 2012 and was no longer active, but later reports would confirm the program continued to be active past 2017, and had been renamed the "Unidentified Aerial Phenomenon Task Force" (UAPTF). This was confirmed by AATIP's retired director, Luis Elizondo, a prior active-duty military intelligence officer who had been recruited into these secretive organizations to investigate these reported unknown objects, and determine if they posed a threat to National Security.

To continue to add to confusion, the UAPTF is now known as "All-domain Anomaly Resolution Office" (AARO) was created in 2022 to replace the UAPTF, and also is located within the Office of the Secretary of Defense. The mission statement for the public is AARO investigates Unidentified Flying Objects and other phenomena referred to as Unidentified Aerial Phenomena (UAP) that may occur on land, the air, the sea, and space.

Additionally in October of 2022, NASA established an independent study team to focus on UAP. In September of 2023 this team released a report stating they "did not find any evidence that extra-terrestrial life was responsible for the unexplained phenomena of UAP sightings".

It seemed odd that NASA, in 2022, would establish an independent team focused on studying Unidentified Aerial Phenomena, since NASA was established in 1958, and many reports from astronauts that they had witnessed extraterrestrials there must have been investigations into those sightings prior to 2022.

With over 77 years of government organizations collecting information and investigations of Unidentified Flying Objects and other related Unidentified Phenomena, it seems this is a matter the Department of Defense takes seriously and collection of data and investigations continues to be conducted mostly in secret. Can everything continue to be attributed to weather balloons and swamp gas?

Will organizations such as AATIP, UAPTF, or now AARO reveal what these unidentified flying objects are if they are not a threat to our National Security. I don't expect any government organization to reveal what they know, partly due to the Army, Navy, and Air Force Joint Regulation 146, which could put someone in jail and possibly fine them a huge sum of money along with confinement for discussing classified UFO reports, as well as the non-disclosure statements that have been signed by anyone with knowledge of this issue which prevents people from sharing what they know.

But what about private civilian organizations set up to investigate UFOs and UAP... will they share the information with other government agencies, will they investigate what is submitted to them, and if they do investigate what they receive what happens if they determine the object is an unidentified phenomenon, or a secret military project, will they inform you of the findings or "ghost" you like you never existed?

Speaking from personal experience this is exactly what I experienced in 2020 after being introduced to an individual associated with a private scientific organization. My pictures, and video, along with details of my sighting were requested and sent to them. I received confirmation they were received, and in further conversations I was told:

- "I have the book and am getting through it. Did you reach out to Tyndall to see if they had a radar corner reflector up that day or get anything from them?" *The book mentioned is Hiding In Plain Sight.*

In further communications I received the following:

- "I have been able to zoom in up to 2400 magnification and can evaluate pixels.", "There is a lot to do and it will take time. I am going to create stacks where I can match up clouds and analyze for motion. The video was helpful."

Communication ended after that, ghosted like I never existed, did they find something they were forbidden to speak about, was my sighting of an unidentified flying object subject to Joint Regulation 146, or was it categorized within the area above Top Secret, and exposure would confirm what the government wants to keep secret?

So, why report sightings if nothing is ever revealed after a report is submitted, unless to say it is a weather balloon or an anomaly with the camera. What answers are there for witnesses, or victims that have been abducted or injured from being in close proximity to the observed objects.

Based on lack of transparency it would appear the reports of Unidentified Flying Objects are not taken seriously, except by intelligence organizations that will always be interested in what is being reported, and of course the USAF that we know will shoot them down.

| **four** |

The Bottom Line

Volume II does not bring new information that identifies what I took pictures of, but what it does bring is a new presentation of the pictures that have been enhanced through the use of filters and software, thereby exposing details not previously seen, just like you would expect any scientific organization to conduct in order to find answers.

The individual I was introduced to, who also performed the process to enhance these pictures, had the following to say about them:

- *"These pictures are not the usual shape (spheres and saucers) most folks can mentally process and will be hard to digest. My point is, the true off-world stuff is hard for most to latch onto. Life is massive and weird. Until your paradigm changes to handle it, it remains an emotionally significant event."*

- *"I personally extract UFO-type anomalies from solar satellite images. Therefore, I am very familiar with seeing shapes no one else has ever seen and come to the same conclusion as you that these are not natural and powered by intelligence."*

It is possible the Department of Defense has developed, or acquired the technology captured by my camera; technology that is so far advanced it has never been seen. I don't believe the answer I received in 2019, that what I took pictures of is a balloon or that it is due to an anomaly with the camera. Nor do I believe it is a kite, or a sheet of mylar floating around the skies over Florida. This object was under control by intelligence, possibly Non-Human Intelligence, that was either part of it, or remotely controlled from another location. So, what are the options for what it is?

In my mind, it isn't a question, I believe 100 percent it is real. I observed this object for approximately three hours, the pictures and video I took are real, but the real question is: if it is not a project controlled by the United States Department of Defense, then what is it, who owns it, and where did it come from?

The object was not static, it was in continuous motion, it continued to change its shape, the cycles seemed to be determined with a purpose, rather than a random series of events without an end goal; a "metamorphosis".

The object was able to stay aloft without any visible means of propulsion that resembled modern aircraft, rather it is fixed or rotary wing. It was able to move at high speeds both horizontally and vertically without regard to gravity, and possessed the ability to manipulate light to conceal itself. There isn't anything else like it, if there is... where is it?

The Department of Defense through contracts with private industry builds classified aircraft and often those highly secret craft are seen and reported as an Unidentified Flying Object (UFO), so it makes sense for anyone with knowledge of these programs to deny they know what they are, or where they come from. Basically, it becomes a built-in cover story to maintain secrecy for those crafts, but not everything observed in the atmosphere comes from human ingenuity and many sightings remain unidentified.

In recent years since 2017 there have been videos and pictures released that show unidentified objects. Several videos taken by military pilots of unidentified objects have been confirmed to be real and authentic videos by the US Navy. The unidentified objects in the videos have been seen traveling in the air and water at extremely high rates of speed.

There have been many statements from government officials that they do not know what these unidentified objects are, nor do they belong to the United States. They also claim these unidentified do not belong to China or Russia; they do not know where they come from. Some wonder if they are a threat to National Security. If these unidentified objects are as technologically advanced as some claim, then it may not be just National Defense that is a concern.

Just a few of the comments from high-ranking government officials, and "whistleblowers" that claim to have knowledge of these unidentified objects include:
Former President Barack Obama:

- "What is true, and I'm actually being serious here, is that there is footage and records of objects in the skies that we don't know exactly what they are,"

Former director of national intelligence, John Ratcliffe:

- "There are a lot more sightings than have been made public," "Some of those have been declassified. And when we talk about sightings, we are talking about objects that have been seen by Navy or Air Force pilots, or have been picked up by satellite imagery that frankly engage in actions that are difficult to explain. Movements that are hard to replicate that we don't have the technology for. Or traveling at speeds that exceed the sound barrier without a sonic boom."

National Security Council Coordinator John Kirby:

- "We obviously take the issue of unidentified aerial phenomena seriously," he said. "There's a whole office at the Pentagon that is stood up to analyze the data, collect reports, collate those reports, and forward them up appropriately."

- "What we believe is that there are unexplained aerial phenomena that have been cited and reported by pilots, Navy and Air Force, and that these phenomena have in some cases had an impact on our pilots' ability to fly, train, operate, and stay ready," Kirby said. "That alone makes it a national security issue worth looking at."

David Grusch, Major, USAF (Ret), (Whistleblower)

- Testified during a House of Representatives hearing, that the U.S. is concealing a long-standing program that retrieves and reverse engineers unidentified flying objects or UAPs. *The Pentagon has denied his claims.*

Former President Ronald Regan, why did he include a statement about an alien threat:

- "In our obsession with antagonisms of the moment, we often forget how much unites the members of humanity. Perhaps we need some outside universal threat to make us recognize this common bond. I occasionally think how quickly our differences worldwide would vanish if we were facing an alien threat from outside this world and yet I ask you, is not an alien force already among us? What could be more alien to the universal aspirations of our peoples than war and the threat of war?"
-

President Regan may also have been the first President to take action to protect the United States as well as the planet with his Strategic Defense Initiative (SDI), which became known as Star Wars Program due to parts of the defensive system would be located in space. SDI was intended to defend the United States from intercontinental ballistic missiles fired from at the time, the Soviet Union. SDI would be composed of both space and earth-based laser systems capable of shooting down incoming moving targets from space.

Could this system be used in defense if attacked by something other than ballistic missiles? Maybe President Regan knew something that prompted him to take such actions of building a space-based system. Based on Project Blue book, an official program that were supposedly closed, but years later find out the program continued to exist under a different name; in my opinion, SDI never shut down, but was renamed and possibly moved to under another organization.

What isn't a secret is that if there is an object in the atmosphere over the United States; it could be shot down by military forces. Where the objects shot down in 2023 considered "unidentified"? Was National Security at risk when the United States Air Force shot down unidentified objects in the air over the United States and Canada? In 2023 the United States Air Force shot down at least four objects. Only one was publicly identified as a Chinese spy balloon and shot down off the coast of South Carolina, where it crashed into the Atlantic Ocean, but what were the other three objects? Surely the government did not order the USAF to shoot down its own technology. What were those objects? Were they a threat to the United States? Why hasn't the government provided information on what it shot down?

What about the three unknown objects that were shot down, all the public is aware of is one shot down over Alaska, one shot down over Canada, and another shot down over the Great Lakes, and none were identified as Chines Spy balloons, so what were they? When will the next unidentified flying object be shot down over the United States? With actions like these, and no explanation, it is easy to believe the government is not telling the truth about what these objects are.

Why wasn't the object I saw and took pictures of shot down? Could it be that it appeared and was not detected on radar… that it possible. What is more concerning: that an object can appear and disappear in controlled airspace and not be detected by the systems in place to protect us, or was this something the government was aware of?

Truth may not come in one press release but it is always there, "breadcrumbs" are ever present. Executive Order (EO) 13526 issued 29 Dec 2009 by President Obama, was put in place so previously classified information could become declassified systematically as soon as practicable Executive Order 13526 may have provided a glimpse of truth by allowing the FBI to declassify and release documents classified for over fifty years, specifically "Memo 6751", ("A Memorandum of Importance") a report prepared in July 1947 by an FBI special Agent. The writer of this memorandum remains unknown, but suspected to hold several degrees and was at some time a department head of a University. What is significant about the report is, that it became a classified document.

What is very interesting about Memo 6751 is how it aligns with Hinduism scriptures. There isn't mention of extraterrestrial beings, rather Hindu Scriptures teach there are multiple universes, as well as multiple dimensions full of life with intelligent beings and advanced civilizations. The Lokas are worlds, spheres or localities, and Talas represent planes of consciousness.

THE ROUND ROBIN

THE FLYING ROLL

San Diego, California July 8, 1947 -

(For your Information)   - A MEMORANDUM OF IMPORTANCE -

THIS MEMORANDUM is respectfully addressed to certain scientists of distincti
to important aeronautical and military authorities, to a number of public officia
and to a few publications.

The writer has little expectation that anything of import will be accomplish
by this gesture. The mere fact that the data herein were obtained by so-called
supernormal means is probably sufficient to insure its disregard by nearly all th
persons addressed; nevertheless it seems a public duty to make it available. (The
present writer has several university degrees and was formerly a university depar
ment head).

A very serious situation may develop at any time with regard to the "flying
saucers." If one of these should be attacked, the attacking plane will almost c
tainly be destroyed. In the public mind this might create near panic and internat
ional suspicion. The principal data concerning these craft is now at hand and mus
be offered, no matter how fantastic and unintelligible it may seem to minds not
previously instructed in thinking of this type.

1. Part of the disks carry crews, others are under remote control.
2. Their mission is peaceful. The visitors contemplate settling on this pl
3. These visitors are human-like but much larger in size.

4. They are NOT excarnate earth people, but come from their own world.
5. They do NOT come from any "planet" as we use the word, but from an other
    planet which interpenetrates with our own and is not perceptible to us.
6. The bodies of the visitors, and the craft also, automatically "materiali
    on entering the vibratory rate of our dense matter. (Cp. "apports.")

7. The disks possess a type of radiant energy, or a ray, which will easil
    disintegrate any attacking ship. They reenter the etheric at will, and
    so simply disappear from our vision, without trace.
8. The region from which they come is NOT the "astral plane", but corres -
    ponds to the Lokas or Talas. Students of esoteric matters will underst
    these terms.
9. They probably cannot be reached by radio, but probably can be by radar,
    a signal system can be devised for that apparatus.

We give information and warning, and can do no more. Let the newcomers be
treated with every kindness. Unless the disks are withdrawn, a situation may arise
with which our culture and science are incapable of dealing. A heavy responsibil
ity rests upon the few in authority who are able to understand the matter.

San Diego, California.

Addendum: The Lokas are oval shape, flute length
oval with a heat-resisting metal or alloy not yet k
the front edge contains the controls; the middle portio
laboratory; the rear contains armament, which consists essentially of a powerful
energy apparatus, perhaps a ray

If it is a Department of Defense program, then it is in the best interest of National Security for intelligence agencies to continue to promote the theory that there are Unidentified Flying Objects and other unknown phenomena in the atmosphere and possibly in the oceans or within the planet, that they don't know what it is, or where it came from.

If it is something that is non-human in nature, either extraterrestrial or biological, that too would most likely be something the Department of Defense would want to keep secret. However, keeping it a secret may not be possible when no one knows what it is or where it came from, the Unidentified Phenomena may make the choice when they decide the time is right.

The Unidentified Flying Object, the "Flying Saucer", may be the ultimate secret or it continues to be the ultimate Deception and Concealment program.

The Bottom Line:

The truth has not revealed itself... yet.

ABOUT THE AUTHOR

Scott Gearen was born in Florida and grew up in the Tampa Bay Area, at a time when it seemed like there were more cows in Florida than people, you could still see the beach from the road, Orlando was not a tourist attraction, Saturn 5 was taking astronauts to the moon, 10-pound bass at Fisheating Creek were not rare, and Alligator Alley was a two-lane highway without fences to keep the alligators off the road.

In 1979 Scott joined the United States Air Force with the goal of becoming a Pararescueman, often referred to as a "PJ". The term PJ was derived from the two-letter designator code needed for aircrews to identify their positions, the call sign stuck and Pararescuemen were forever referred to as PJs. After successfully passing the Pararescue Selection course, his first military assignment was to Royal Air Force Base Woodbridge in 1980 where stories of Unidentified Flying Objects were on everyone's lips after the Rendlesham Forest incident.

After 22 years of active duty and adventures around the world he was medically discharged from the Air Force. Although retired from the Air Force the desire for lifetime adventures was not retired, and as a founding member of the Old Man Adventure Club, he continues to enjoy trips around the world, and always with a camera, because you never know when you will see something that no one is going to believe without the pictures to prove it...

ACKNOWLEDGMENTS

Sabrina Robb: I don't think it was a coincidence that you were the Field Investigator assigned to my case, and I don't believe anyone else would have come close to what you provided through your investigative skills, first hand experiences, vast network in the UFO community, ICBM launch officer, and stories still not talked about from silos in the Midwest to airfield encounters at Charleston AFB. Thank you for all the conversations discussing the unproven theories and hypothesis about the things we know to be true.

Anonymous: Thank you for your service and personal sacrifices in life, in government service, and your continued efforts to uncover the truth. Your expertise in media has made it possible to see what was hidden. Through our conversations I have a greater understanding of the emotional tear after my encounter and rebuilding of a new paradigm. Expect lots of growth ahead due to the inevitable paradigm shifts as the truth is uncovered.

Steph @covertress: Bowline to the "Sky-Ray" and its clear sailing.

L. @Sh_tNoFake: Definitely a "space critter".

L. @just_yourghost: For keeping it front and center.

www.ingramcontent.com/pod-product-compliance
Lightning Source LLC
Chambersburg PA
CBHW042045130726

48010CB00027B/281